Mathematics of the Incas:
Code of the Quipu

Marcia Ascher
Department of Mathematics
Ithaca College

Robert Ascher
Department of Anthropology
Cornell University

DOVER PUBLICATIONS, INC.
Mineola, New York

To our parents
and the people
of the book

Bibliographical Note

This Dover edition, first published in 1997, is an unabridged and corrected republication of the work first published by The University of Michigan Press, Ann Arbor, in 1981 under the title *Code of the Quipu: A Study in Media, Mathematics, and Culture.*

Library of Congress Cataloging-in-Publication Data

Ascher, Marcia, 1935–
    [Code of the quipu]
    Mathematics of the Incas : code of the quipu / Marcia Ascher, Robert Ascher.
       p.   cm.
    Originally published: Code of the quipu. Ann Arbor : University of Michigan Press, c1981.
    Includes bibliographical references.
    ISBN 0-486-29554-0 (pbk.)
    1. Quipu. 2. Incas—Mathematics. 3. Indians of South America—Andes Region—Mathematics. I. Ascher, Robert, 1931–
II. Title.
F3429.3.Q6A82   1997   96-47494
302.2'22—dc21   CIP

Manufactured in the United States of America
Dover Publications, Inc., 31 East 2nd Street, Mineola, N.Y. 11501

Preface

This book is about an artifact that became the medium for expression in a major civilization. The medium, its content and its context—the quipu in the Inca state—is our focus. But we see our work in a broader frame. There are limitations on anyone's capacity to understand what is going on in another contemporary culture. The limitations are compounded when trying to understand what went on in a culture that was destroyed centuries ago. The effort forced us to question many previously held notions. In particular, we had to rethink some of our ideas about mathematics, media, and culture. And we were forced to reexamine how we and others codify the universe that surrounds us. We learned from the effort, and here, for the reader, we convey what we learned.

The book does not assume any specialized knowledge. It is intended for anyone who is curious and willing to read participatively. Exercises and discussive answers are included in some of the chapters. Notes are keyed to chapter sections. They include source references and places for further reading about a topic, as well as some comments that we wanted to make but are tangential to the flow of the text. We do not describe all quipus in all their details. However, the book gives a very good sense of them. There are photographs of eight quipus; thirty are explicitly discussed in the context of ideas; and four are used in the exercises.

Our interest in the problems posed by quipus started about ten years ago. At first, we worked from quipu descriptions published by others. Then, we started to examine specimens close to home. Our early work taught us a few things. The colored cotton or wool strings from which quipus were made could decompose even under good conditions. (We had the experience of touching a quipu string only to withdraw quickly as it collapsed into a small heap of dust. Fortunately, all quipus are not in such poor condition.) The published descriptions were too few and too variable to permit us to follow up some of our initial ideas. So, prior to further attempts at interpretation, we set out to locate, and to describe in a uniform fashion, as many quipus as we could. We eventually located specimens in museums and private collections in South America, Europe, and North America. We asked permission to study and describe them, and we received it in all but one instance.

We are indebted to those people who permitted us to study and describe the quipus that are the primary basis for our study. They are: Jorge C. Muelle, Edward Versteylen, Alicia Gamarra, Ana María Soldi, and Yoshitaro Amano (Lima, Peru); Oscar Núñez del Prado (Cuzco, Peru); A. Pezzia (Ica, Peru); Percy Dauelsberg (Arica, Chile); José María Vargas (Quito, Ecuador); Mireille Simon-Abbat (Paris, France); Daniel

Schoepf and A. Jeanneret (Geneva, Switzerland); Gerhard Baer (Basel, Switzerland); Jacques J. Buffort (Rotterdam, Netherlands); Hans Becher (Hanover, Germany); Imina von Schuler and Dieter Eisleb (Berlin, Germany); Otto Zerries (Munich, Germany); J. M. Pugh (London, England); Michael Kan, Frederick Docksteader, Junius Bird, and Craig Morris (New York, New York); Clifford Evans (Washington, D.C.); Thomas C. Greaves (Philadelphia, Pennsylvania); Frank A. Norick and Lawrence E. Dawson (Berkeley, California); Christopher B. Donnan (Los Angeles, California); and Gordon R. Willey (Cambridge, Massachusetts). The list is a testimony to international cooperation in scholarship.

In most cases, we had to travel to where the quipus are stored. The Wenner-Gren Foundation for Anthropological Research provided partial assistance for our foreign journey. We thank the foundation, and in particular its director, Lita Osmundsen.

The descriptions we recorded covered more than a thousand typed pages. Carol Mitchell of the University of Michigan Press understood the importance of making primary information available to any interested person. The descriptions are now published (see the Authors' Note). We thank Ms. Mitchell for her understanding.

In the course of the decade of intermittant work, many people provided assistance of one kind or another. We gratefully acknowledge their help. They are: David H. Abrahams, William B. Bergmark, Marjorie Ciaschi, Denise Everhart, Ann Fairchild, Steven Frankel, Stephen Hilbert, Billie-Jean Isbell, Ruth H. Johnson, Lawrence Kaplan, Margaret LeBoffe, Mercedes López-Baralt, Duncan Mason, Dorothy Menzel, Elias Mujica, Gerald Munchel, John V. Murra, Joan Oltz, Dottie Owens, William Pinkham, John H. Rowe, Mary Saintangelo, Max Saltzman, Steven Sayardar, D. J. Struik, Gary Wayne, and R. L. Wilder.

We appreciate the use of the photographs and figures that are not our own. These are: plate 2.4, Department of Anthropology, Smithsonian National Museum, Washington, D.C.; plates 3.2, 5.1, 5.2, 5.3, and 5.4, Museum für Völkerkunde, Berlin; plate 3.6, and our redrawn plan of part of Inty Pata (plate 3.7), are from Paul Fejos, *Archaeological Explorations in the Cordillera Vilcambamba Southeastern Peru*, Viking Fund Publications in Anthropology, No. 3, Copyright 1944 by the Wenner-Gren Foundation for Anthropological Research, Inc.; plates 3.3 and 3.4, Institute of Andean Research, and American Museum of Natural History, respectively; plate 3.1, Fratelli Alinari, Florence, Italy; plate 4.1 is from J. B. Nies, *Ur Dynasty Tablets*, 1919, J. C. Hinrichs'sche Buchhandlung, Leipzig; and plates 4.2 and 4.3 are from Felipe Guamán

Poma de Ayala, *Nueva Crónica y Buen Gobierno,* Institut d'Ethnologie, Université de Paris.

We hope that you enjoy the book.

Seal Harbor, Maine
1979

AUTHORS' NOTE

The interpretations in this book draw primarily upon our *Code of the Quipu Databook.* It was published in 1978 by the University of Michigan Press and is available on microfiche from Cornell University Archives, Ithaca, New York.

The databook contains descriptions of 191 quipus. These were obtained by first-hand study of specimens in museums and private collections spread over three continents. The introduction includes information needed to understand the databook descriptions, the locations of all known quipus, and a complete bibliography of previously published descriptions. The descriptions in the databook are written in such a way, and in sufficient detail, so that any interested person can formulate his or her independent interpretations.

Contents

Chapter **1** | # Odyssey

1 *One of the three brothers had a golden sling, and with it he could throw a stone up to the sky: it would almost touch the clouds.*

These are words from a story that the Incas told about their own origins. At first, it seemed a good way to start: what better way is there to begin to discuss the quipus of the Incas than by using an Inca story about their own beginnings. But right away we must pause; the story was spoken in Quechua, recorded in Spanish, and translated into American English. Once aware of this, we cannot go further until we answer the question, How do we know anything about the Incas?

The Spanish recorded the Inca origin story more than four and a half centuries ago. The Incas were a culture, a civilization, and a state. That is to say, the word Inca, as we use it, applies to particular forms of human association. The land that the Incas once occupied is today all of Peru and portions of Ecuador, Bolivia, Chile, and Argentina. When the Spanish arrived to conquer, the Inca state had existed for about one

1 There are several versions of the Inca origin story. Parts of different versions appear in Harold Osborne, *South American Mythology,* (Feltham, Eng.: Hamlyn Publishing Group, 1968).

hundred years. Within thirty years—the number of years generally used to designate one human generation—Inca civilization was destroyed.

The Incas did not write as we usually understand that activity. For written accounts of Inca culture, we must turn to the sixteenth-century Spanish of soldiers, priests, and administrators. Yet, the culture of the Spaniards of that time is remote from our own. We do not share with them, for example, a real fear of the devil, even if we are part of the tradition that invented him. And the devil, together with many other cultural predispositions, figured largely in Spanish discussions of the Incas. To make matters worse, the Spanish got their information almost exclusively from deposed Inca bureaucrats. They were a special and numerically small part of a population estimated at somewhere between three and five million people. Whatever we may or may not have in common with sixteenth-century Spaniards, they shared close to nothing with the Incas. We can make sense out of Spanish accounts only in terms of our framework, and the Spanish, for their part, rendered what the Incas said from inside a Spanish framework. As a result, written accounts are distorted as they pass through this route: one culture (Inca) is interpreted via a second culture (Spanish), which is interpreted via a third culture (American), four hundred and fifty years later.

Luckily, there is a source of knowledge in addition to writing. Walking one day in the streets of Cuzco, once the capital of the Inca state, we saw an Inca wall, topped by a Spanish wall, on which was hung a Coca Cola sign. What we saw tells a good deal about the relationship between three cultures. But let us concentrate on the Inca wall. The Spanish could hear about the wall, and they could see it and touch it. We also can see it and touch it, but we cannot hear about it from the Incas. Nevertheless, the Inca wall, and other things that they made and that have survived, provide us with a direct way of knowing about the Incas.

Using material things as a source of knowledge does not, however, do away with distortion. Walls of some sort occur in every culture. This can lead us to think that wherever they occur they have the same meaning. They do not. Nevertheless, with due caution, it is reasonable to assume that walls, wherever they are found, serve a roughly similar purpose. But there is another more difficult problem in understanding material evidence. There are some things in one culture for which there are no counterparts elsewhere. When this happens, understanding becomes even more difficult for someone outside the culture. For example, native Australians had no counterpart of the airplane. It was at first difficult for them to understand that people flew through the air in metal containers. And the problem increases when an attempt is made to know about a culture that is remote in time as well as in space.

2

We wish to understand the quipus made by the Incas. But unlike walls, there were no counterparts of quipus in sixteenth-century Spanish culture and there are none in our own experience. A quipu is a collection of cords with knots tied in them. The cords were usually made of cotton, and they were often dyed one or more colors. When held in the hands, a quipu is unimpressive; surely, in our culture, it might be mistaken for a tangled old mop. For the Spanish, the Inca quipu was the equivalent of the Western airplane for native Australians. For us, the problem is compounded by a separation of four and a half centuries. Before rushing ahead to where touching, seeing, and thinking about quipus led us, a context for them must be provided. To do this, we call upon a Spanish witness.

2 Spanish writers shared a cultural framework, but there were very important individual differences. Today there is general agreement on which of the writers are relatively reliable. Cieza de León was perhaps the most reliable. He was a good observer and a careful listener, and he was the first person to write about quipus. Our knowledge of what the Spanish understood about quipus is increased only a little by going to other early writings, for Cieza understood more than his contemporaries, and many later writers simply copied from him.

The writings of Cieza seized and held our attention from the start of our studies. There are minor reasons and one major reason for this. Cieza was in Inca territory only fifteen years after the conquest; this alone commends his work. He saw things that others who followed cannot have seen, and he spoke with people who were adults at the apogee of Inca power. In addition, Cieza is an able writer. His choice of vocabulary, his ability to put things concisely, his conscientious attempt to weigh evidence, and his respect for the intelligence of the reader, set him apart as much as does his being there before others.

But the major appeal of Cieza centers on what leads us to use the word odyssey to describe his work. At one level, an odyssey simply means a journey; that clearly applies to Cieza. In a broader sense, an odyssey is a quest or a search that may or may not involve travel through real space. It is important that the term odyssey in this second sense also applies to Cieza. In literature, the odyssey is often a self-conscious search for the self. Cieza's quest is different: he is not looking for himself, but rather for the substance of the victims of conquest. That Cieza thought of the Incas as victims is clear; as he says, "wherever the Spanish have passed, conquering and discovering, it is as though a fire had gone, destroying everything it passed." Although he was a party to conquest, the victor is of less interest to him than the vanquished.

2 The first part of the writings of Pedro Cieza de León was published in 1553, the second part in 1880, and the third part was published as late as 1946. The best English translation of the first two parts is by Harriet de Onís. It is in Victor Wolfgang von Hagen, ed., *The Incas of Pedro de Cieza de León* (Norman: University of Oklahoma Press, 1959). Our English citations from the works of Cieza derive largely from the Harriet de Onís translation, but Spanish versions have been consulted where there is a question of interpretation. There are two major sources on Cieza's life. They are: Marcos Jimenéz de la Espada, *Prólogo de la Segunda Parte de la Crónica del Perú de Cieza de León*, vol. 5 (Madrid: Biblioteca Hispano-Ultramarina, 1880); and Miguel Maticorena Estrada, "Cieza de León en Sevilla y su Muerte en 1554," *Anuario de Estudios Americanos* 12(1955): 615–73. In his editor's introduction to the Harriet de Onís translation, von Hagen provides a background of Cieza's times and a summary of his life. A study of the literary aspects of Cieza's writings is found in Pedro R. León, *Algunas Observaciones Sobre Pedro de Cieza de León y la Crónica del Peru* (Madrid: Biblioteca Románica Hispánica, 1973). This book also contains a rather complete bibliography on the work and life work of Cieza.

For English language readers who want to pursue what chroniclers other than Cieza had to say about quipus, the best place is the excerpts section at the back of Leland L. Locke, *The Ancient Quipu or Peruvian Knot Record* (New York: American Museum of Natural History, 1923). This section contains translations from about fifteen early Spanish sources. They vary in length from a few lines to several hundred lines of print. Some other sources have come to light since Locke wrote. See, for example, H. Trimborn, *Quellen zur Kulturgeschichte des Präkolombischen Amerika* (Stuttgart: Strecker und Schröder Verlag, 1936). John V. Murra, "An Aymara Kingdom in 1567," *Ethnohistory* 15 (1968): 115–51, presents data that is said to have been originally recorded on a quipu; our discussion of this possible quipu is in "Numbers and Relations from Ancient Andean Quipus," *Archive for the History of Exact Sciences* 8 (1972): 288–320.

In most odysseys, the route is selected by the traveler. Not in this case. Cieza was a common soldier, and as such he was more or less told when and where to move. The latter part of his journey interests us. It began in April, 1547, when he walked into the northern extreme of the area once ruled by the Incas. Cieza was then twenty-nine years old. He moved along good roads, ones which, he says, were superior to the Roman roads he knew as a boy in Spain. Day after day, however wearied, he paused and recorded some of the things he heard and saw. He stopped writing in September, 1550.

As we follow Cieza on his quest for the Incas, keep in mind that he lived over four centuries ago and that he describes something that was very different from what he knew. We convey the sense of his vision as we understand it. Even under the best circumstances, another culture is blurred as if seen through heavy gauze. Or, as Cieza put it, "Peru and the rest of the Indies are so many leagues from Spain and there are so many seas between."

And as the traveler trudges through all this sand and glimpses the valley even though from afar, his heart rejoices especially if he is traveling on foot and the sun is high and he is thirsty.

In the morning, Cieza set out from a place where he was soaking wet; by nightfall, he was in a region where it never rained. To do this, he traveled west and down from the mountains and then stopped when he reached the desert sands. In the late afternoon of another day he went south through the desert. Alternately, but still heading south, he stayed with mountains or the plains. The roads that he followed can be visualized as a ladder resting unevenly upon the ground. One side of the ladder runs through the mountains and is raised; the lower side goes through the desert at the edge of the Pacific Ocean. The crossbars of the ladder go down through the valleys connecting the mountains with the desert and the sea. At its greatest, a side of the ladder, as Cieza counted it, was 1,200 leagues (about 3,600 miles). The crossbars varied from as little as one hundred twenty miles to as long as three hundred miles. The roads, built by the Incas, roughly defined the extent of their rule.

They observed the customs of their own people and dressed after the fashion of their own land, so that if there were a hundred thousand men, they could easily be recognized by the insignia they wore about their heads.

4

Cieza, the quester, was impressed by the diversity he found as he moved along the road. He noticed changes in animals, crops, weather, and landscape. Mostly, he was struck by the differences in groups of people. He might anticipate some of these: for example, people in rainy mountainous regions usually built houses of fieldstone; in the desert, clay bricks dried in the sun were used for the same purpose. There were also marked differences in customs, languages, myths, and sexual practices. Regarding the latter, Cieza often inquired about the diabolical sin of sodomy; he was very disturbed when he was told that it existed in one place or another. Beyond these observations, Cieza noted that people differed in the directness of their speech, their looks, and their attitudes. The quality was somehow different. For example, two groups living in the mountains might prefer wool for making clothing, but the bearing of the people, and hence the way their clothing fell about them, was not the same in both groups. Originally, the Incas were a group of people living in the mountains to the south. The ruling Inca family, the head of which we call Sapa Inca, started with three mythical brothers, whose thoughts, according to Cieza, soared high. The Incas differed from their neighbors just as their neighbors differed from others living close by. In our terms, the Incas were a culture, and each of the other groups were separate, more or less self-contained cultures.

He sent messengers to these people with great gifts urging them not to fight him for he wanted only peace with honorable conditions and they would always find help in him as they had in his father and he wished to take nothing from them but to give them what he brought.

At almost every step along the way, Cieza noted the presence of one culture in the midst of all others. There was no puzzle in this. Going step by step, and increasing the pace in the three generations before Cieza, the Incas moved upon their neighbors. Doing this, they upset the solitude of the cultures in western South America. The Incas moved upon a group as if they were the bearers of important gifts. A deity bringing better times, or a method to make the land more fruitful, or food if the need for it existed, are examples of the gifts. If the gifts were accepted, there was no need for violence; if not, force was applied. In any case, the gifts were delivered, and thus, selected parts of Inca culture were everywhere superimposed. According to Cieza, the Incas did not want to destroy and replace the cultures already there. For example, an Inca deity was added to, not substituted for, the local gods. And locally important people continued to be important, even if they now had to take an interest in what the Incas wanted done.

There were even provinces where, when the natives alleged that they were unable to pay their tribute, the Inca ordered that each inhabitant should be obliged to turn in every four months a large quill full of live lice, which was the Inca's way of teaching and accustoming them to pay tribute.

Cieza did not go far in any direction before he came upon compelling physical reminders of Inca rule. This took the form of a complex of buildings that were put up soon after the Incas took over. The components were mostly the same, regardless of region. There was a temple to the sun, storehouses, and lodgings. The buildings were now empty, their furnishings removed, and their elaborate decorations torn away. The bureaucratic officials, religious functionaries, and military personnel who peopled the buildings were gone, of course. When the buildings were in use, control had come from Cuzco, the principal Inca city, but it was hundreds of miles away, and particulars had to be administered locally. For example, tribute was fixed in Cuzco after an investigation to determine the form the tribute should take. The collection of tribute, however, was in the hands of local people. Another function of regional supervision was associated with the Inca policy of moving, en masse, thousands of people from one place to another. After the colonists arrived at their new place, it was the people in the Inca buildings who arranged the details of resettlement. Cieza thought that the tribute was levied fairly. He also thought well of the removal policy, pointing out that the colonists were often taught new trades.

When the great dances were held the square of Cuzco was roped off with a cable of gold which he had ordered made of the stores of the metal which the regions paid as tribute, whose size I have already told, and an even greater display of statues and relics.

"I recall," writes Cieza, "that when I was in Cuzco last year, 1550, in the month of August, after they had harvested their crops, the Indians and their wives entered the city making a great noise, carrying their plows in their hands, and straw and corn, to hold a feast that was only singing and relating how in the past they used to celebrate festivals." In the past and during Cieza's visit, festivals were held in the central square of the city. Literally and figuratively, the four highways which partitioned the world of the Incas into four main divisions met at the center of the square. In the same square, the men who held leadership powers were confirmed in their authority amid theatrical displays of their inheritance rights. The remains of their ancestors were paraded about. The Incas founded Cuzco, resided in it, and ruled from it; yet, the people in the city included non-Incas. The crowd at festivals was made up of men from other cultures dressed in their traditional garb. They were there to keep up the connection between the Incas and the people whom they ruled. And they were there to learn the ways of the Incas. Cieza thought that Cuzco was a glorious place, with large buildings, wide streets, and lavish displays of wealth. It had, he says, "an air of nobility." He was just as certain that with so many different people in Cuzco, wizards, witches, and idolators were plentiful, and the devil, having insinuated himself, held sway.

Not a day went by that posts did not arrive, not one or a few but many, from Cuzco, the Colla, Chile, and all his kingdom.

The experienced traveler, and that Cieza was, is not gullible. He preferred to omit a story if it was not reinforced by the testimony of several people, or better yet, by his own observations. In the case of the communication system devised by the Incas for their own use, Cieza writes: "Nowhere in the world does one read that such an invention existed . . ."; so he was cautious. But he found corroboration for what he heard in the shape of small posthouses along the highways. The system was simple. A runner carried a message from the posthouse where he was stationed to the next posthouse up the road. As he approached, he called out for attention and then passed the message on; a new runner took the message the next few miles. This was repeated by successive runners until the message reached its destination. With a certainty unusual for him, Cieza says that the messenger system was superior to one dependent on horses and mules, animals he was familiar with. Only runners could negotiate the perilous mountain passes, suspension bridges, and rocky wastes overgrown with briars and thorns. Through this system, simple though it was, the Inca bureaucracy continuously monitored the areas under its control. The day of a Sapa Inca, according to Cieza, was taken up in large part with receiving messages and sending instructions. The Incas thought of something else to do about communication. They insisted that everyone learn Quechua, the Inca language, along with their native language. In time, many people could speak Quechua, and official business was carried on in that language.

There have been occasions when I stopped beside one of the canals, and before I had time to pitch my tent, the ditch was dry and the water had been diverted elsewhere.

The lowly potato was first described in writing by Cieza. Potatoes are a basic food, akin to rice and wheat. For the benefit of his European readers, he explained that potatoes were like truffles, a food they knew. On his trip into the southern Andes, Cieza found that it was common practice to dry potatoes in the sun, put them away in that condition, and eat them the following year. In Cieza's time, the benefactors of this planning were often limited to the people who grew the potatoes. It was different under Inca rule. Recall that storehouses were part of the regional building complex. They held various items, including food. The Incas responded to a food shortage by taking food from a storehouse and sending it to where it was needed, regardless of who grew it. If food was in surplus, it was distributed to the poor and aged. The benefits of already existent planning schemes, such as stored, dried potatoes, were thus extended across ecological and cultural boundaries and reached those who were not party to the plan. The case of the dried potato can stand as a model for Inca planning and control. For example, the model fits the Inca solution to the water problem. As with dried potatoes, the problem was to move a resource from where it was abundant to where it was needed. The solution was irrigation and terracing. Irrigation, like the cultivation and storage of potatoes, predates the Inca expansion. The Incas redeveloped old systems and ordered the construction of new ones. In an agrarian society, nothing is as important as water, except perhaps the sun; and the Incas and the sun had a special relationship if the number of temples dedicated to that star is any guide. Cieza was struck by the beauty the irrigation systems brought: "All this irrigation," he says, "makes it a pleasure to cross the valleys, because it is as though one were walking amidst gardens and cool groves."

> *And little boys, who if you saw them you would not think knew*
> *how to talk yet, understood how to do these things.*

On the slope of a hill overlooking Cuzco, Cieza took the measure of an enormous stone and found that it was 270 hand spans in circumference. It was also very high. Quarry marks on the stone lent credence to the story that it was hauled up the hill to become part of a massive stronghold. The stone that Cieza measured had not reached the top of the hill, but others did, and were there to be seen, as they are today. Large numbers were needed to express the sizes of the stones, the years the project was underway, the dimensions of the overall structure, and the people needed to do the various jobs. Impressed as he was by the scale of Inca projects, Cieza did not neglect to write about the quality of the undertakings. In this instance, he dwelt on the craftsmanship of the masons who, with a few tools, cut and fit the stones so exactly that a coin could not be inserted between them. Skill in dealing with very large things also applied in the small. For example, Cieza wrote about goblets and candelabra made with the use of two or three stones and a few pieces of copper. He admired the cloth made by women using a simple loom. And he applied the term skill to undertakings other than making objects. In a show of statecraft, a Sapa Inca dressed himself in the native garb of each village he moved upon, and in this way, says Cieza, won the people over to his service.

In a word, it was once what no longer is, and by what it is, we can judge what it was.

If a word were chosen to condense the substance of the Incas as Cieza came toward an understanding of it, the word would have to be order. Cieza used the word in connection with almost everything, including the planting of crops, the behavior of armies, and the transfer of power. He went so far as to say that order saved the Incas from total destruction. He elaborated the point this way: when the Spanish went through a region their demands were met with goods from the Inca storehouses; those who paid out the most, and were therefore in trouble, were resupplied from the storehouses of the more fortunate. This

response to Spanish plundering was possible, according to Cieza, because orderly records were kept of what goods had left which storehouses.

The records referred to by Cieza were on quipus. In earlier times, when the Incas moved in upon an area, a census was taken and the results were put on quipus. The output of gold mines, the composition of work forces, the amount and kinds of tribute, the contents of storehouses—down to the last sandal, says Cieza—were all recorded on quipus. At the time of the transfer of power from one Sapa Inca to the next, information stored on quipus was called upon to recount the

accomplishments of the new leader's predecessors. Quipus probably predate the coming to power of the Incas. But under the Incas, they became a part of statecraft. Cieza, who attributed much to the action of kings, concluded his chapter on quipus this way: "Their orderly system in Peru is the work of the Lord-Incas who rule it and in every way brought it so high, as those of us here see from this and other greater things. With this, let us proceed."

His travels at an end, Cieza returned to Spain in 1551 at age thirty-three. A year later, the first part of his odyssey was in the hands of a printer. Another year went by, and after approval by the Holy Order of the Inquisition, the King's Council, and the Council of the Indies, it was finally issued in 500 copies. The next year, 1554, Cieza was dead.

From his last will and testament, we know that at the time of his death, Cieza had in his possession about half of the copies of the first printing of his book. In the portion of the will devoted to the saying of masses, he requests hundreds of them for himself, his family, and others. Included in the list of others are eight masses to be said for an Indian woman called Ana, and another ten for ". . . the souls of Indian men and women in purgatory who came from the lands and places where I travelled in the Indies."

2 | How to Make a Quipu

1 In building a house, making a sweater, or writing a symphony, guiding concepts are formed into specific images which always involve knowledge of the materials to be used. Thus, these creations depend, first of all, on knowing the shapes and stresses that can be taken by bricks and mortar, or knowing about yarn and what it forms when knitting or purling, or knowing about musical notes and the sounds they lead to when interpreted on different instruments. Similarly, the first step toward making a quipu is knowing what materials are used and the significance of different manipulations of the material.

The basic materials are colored cords. The basic manipulations are connecting cords together and tying knots in the individual cords. We begin with the preparation of the cords and how to connect them to each other. Then we discuss the significance of the color of the cords and the significance of the knots that are tied in the individual cords. Elements of the symbolic system, therefore, are introduced along with steps in the construction process. In some instances, the step-by-step construction process is only instructional and not necessarily that followed by an Inca quipumaker.

By the end of the chapter, there will not be anything like a completed house or sweater or symphony. There will be the ideas necessary to design and construct a quipulike object and to express some contemporary messages using elements from the quipumaker's symbolic system. But these are only the basics; they are prerequisite to an understanding of the concepts used by the quipumaker.

2 Individually constructed cords are the basic units of a quipu. The cords are distinctive in that each is at least two-ply and one end is looped while the other is tapered and finished with a small knot.

Fig. 2.1

2 Preparation of Cords

(i) A drop spindle is necessary equipment. It can be created with any straight stick as a shaft and any flat topped round weight as a whorl. A knitting needle and half of a potato or a pencil and half of an apple are ideal. For example, insert a knitting needle through the center of half of a potato.

Fig. N.1

Next, cut a piece of yarn about twice the length of the shaft to be used as a lead. Tie the end of the lead around the shaft just above the whorl and secure it in place by wrapping it around the shaft three or four times. Bring the lead outside and under the whorl; wrap it once around the stem of the shaft; and bring it over the other side of the whorl up to the top of the shaft. Attach the lead about 4 cm. from the top of the shaft with a half-hitch. The spindle and lead are now complete and can be put aside while the yarn to be plied is readied.

Fig. N.2

(ii) To double the ply of yarn, two pieces are twisted together. It is the interlocking of the spurs on the yarn that keep the pieces bound together. Commercially available hanks of wool give better results than commercially available cotton cord. Twisting in a clockwise direction is called a Z-twist and twisting in a counterclockwise direction is called an S-twist. Cut two pieces of yarn of the same length. About 40 cm. is a convenient size with which to start. Test a piece of yarn to see which direction tightens its twist. If it is Z-twisted yarn, use Z-twisting to double the ply; and if it is S-twisted, use S-twisting. Now, place the two pieces of yarn parallel to each other and knot one end of the pair. Tie the lead from the spindle to the pair just above the knot. Hold the other end of the pair firmly in your left hand. Drop the spindle and start twisting the yarn with your right hand just above the top of the shaft.

Fig. N.3

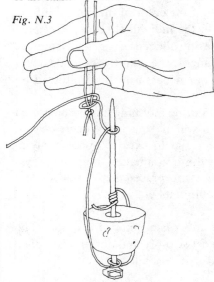

Move your right hand up and down the yarn as you twist. To keep the spindle spinning constantly, occasionally assist the whorl with your right hand. When the pair is tightly twisted together, retaining a hold on the top end, grasp the end connected to the lead with your left hand. Disconnect the lead. (Snipping the yarn and then reknotting the end may be the easiest way.) Knot the top end of the yarn and pull from both ends to straighten.

(iii) For a quipu, each cord has to be at least two-ply and only one end is knotted while the other end is looped. So, as a last step in preparing quipu cords, the ply is doubled but by a slight variation of the method just described. The process uses one piece of yarn twice the intended finished length. The yarn is to be attached to the lead through a small loop. To form the loop in the lead cord, hold the top of the shaft in your left hand; pass the free end of the lead over your index finger; and secure it to the lead just below the half-hitch with another half-hitch.

Fig. N.4

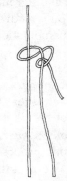

Now, double the yarn to be twisted after inserting one end through the lead's loop. Tighten the loop by pulling the free end of the lead. Keep the end of the lead directed downward so it does not tangle with the yarn as it is twisted. With the pair of free ends of the yarn held firmly in your left hand, drop the spindle and twist as before. When the twisting is done, the lead loop can be opened and the prepared quipu cord slipped off the lead.

(iv) Each quipu usually has one cord (the main cord) that is much thicker than the others. To make thicker cords, the ply of the yarn can be increased using the first method before finally doubling the cords using the modified method.

(v) It is possible to S-twist Z-twisted yarn and vice versa, thus creating differently textured cords. Color variation can be achieved by combining different color yarns by different twistings. If, for example, S-twisted yarns of two different colors are combined by S-twisting, a candy cane pattern results. Doubling this using Z-twisting, gives a cord with a mottled pattern. Later, in the discussion of color, it will be noted that the quipumakers were very inventive in producing many distinctly different looking cords from few basic colors.

The looped ends are used for connecting cords together. When cotton or wool is spun into cord, or if pieces of yarn are cut from longer cords, a special final spinning step is needed to prepare it as a quipu cord. As the last step, the cord has to be bent double and twisted.

With a few such cords at hand, it would be easier to follow the description of how the cords are connected together. But such cords are not familiar or readily available. We suspect that even the Inca quipu-maker finished the cords himself or had them specially finished for him. In the notes for this chapter, there are detailed instructions for preparation of cords by a simple method using commercially available yarn and a drop spindle made from some common items. The method is not descriptive of Inca spinning techniques. It should, however, enable you to produce facsimiles of quipu cords. Carrying out the instructions, or at least reading them through, should also underscore that a quipu is, among other things, a handcrafted item.

3 Individual cords connected together form a quipu. One cord serves as the main cord. Other cords are suspended from it. The pendant cords can be attached as follows:

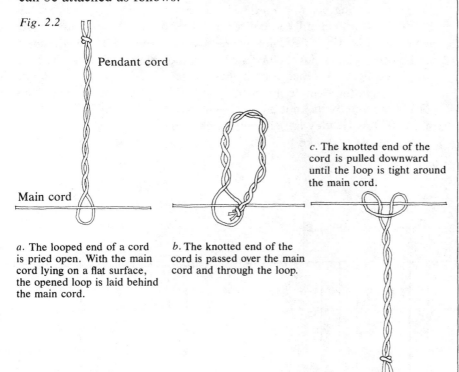

Fig. 2.2

Pendant cord

Main cord

a. The looped end of a cord is pried open. With the main cord lying on a flat surface, the opened loop is laid behind the main cord.

b. The knotted end of the cord is passed over the main cord and through the loop.

c. The knotted end of the cord is pulled downward until the loop is tight around the main cord.

(*vi*) For simple discussions of spinning and knots, see: (*a*) Joseph Leeming, *Fun with String* (New York: Dover Publications, 1974); (*b*) Christine Thresh, *Spinning with a Drop Spindle* (Santa Rosa, Calif.: Thresh Publications, 1971).

3 Detailed descriptions of specific quipus that have the connections discussed in this section can be found in Marcia Ascher and Robert Ascher, *Code of the Quipu Databook* (Ann Arbor: University of Michigan Press, 1978), available on microfiche from University Microfilms International.

Each quipu description in the databook includes the length and color of each cord, the relative positions of the cords, the type of knots in each knot cluster, and the positions of the clusters on the cords. Also included are the current holder of the quipu, the holder's designation for it, and any information on its provenance or acquisition. We use a tag system to refer to published descriptions of quipus. Each quipu is tagged with a letter (or letters) and a number. The letter is derived from the name of the person who published a specimen and the number refers to the order of publication. For example, AS1–AS9 were published by us in "Numbers and Relations from Ancient Andean Quipus," *Archive for History of Exact Sciences* 8 (1972): 288–320. Later, AS10–AS200 were published in the databook. The databook also contains the tags and references for sixty-two quipus published by others.

Top cords of the first type are on, for example, AS10, AS67, and AS115. Top cords of the second type are on, for example, AS44, AS51, and AS199. Examples of quipus with dangle end cords are AS33, AS100, AS103, AS122, AS130, and AS163. Woven balls attached to the main cord are examples of decorative finishings. Examples of these are on AS32, AS57, AS98, and AS195. Some completed quipus, such as AS106, AS124, AS136, and AS140, are suspended from carved wooden bars.

The main cord is generally much thicker than the pendant cords. Most cords are attached to the main cord in the way just described, and so the simplest blank quipu looks like this:

Fig. 2.3

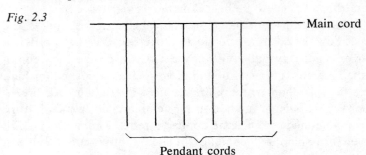

Main cord

Pendant cords

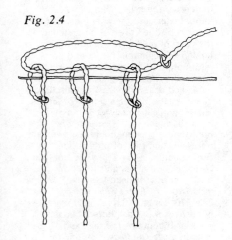

Fig. 2.4

There are, however, a few other ways that cords are attached to the main cord. These will be seen to have special uses. If the knotted ends of the cords are pointed downward when first laid behind the main cord, they point upward when passed over the main cord and pulled tight. Such cords are referred to as *top cords* to distinguish them from the downward facing *pendant cords*. A second way that top cords are attached causes them still to be directed upward but also to unite several pendant cords. For this, some pendant cords are attached to the main cord close together but the last step is not completed. That is, they are not pulled tight. Another cord is then passed, horizontally, looped end first, through the open loops of the pendants and back over them. This cord is secured by its knotted end being inserted through its looped end (fig. 2.4). Lastly, they are all pulled tight.

The results look like this:

Fig. 2.5

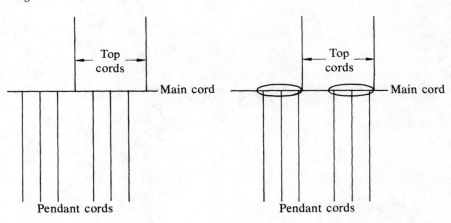

A special cord can also be connected to the end of the main cord. A cord is passed through the looped end of the main cord and then through its own looped end and tightened. The effect is that this cord dangles from the end of the main cord (fig. 2.6).

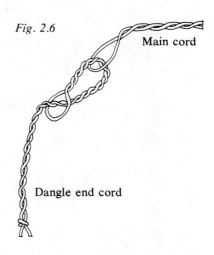

Fig. 2.6

Main cord

Dangle end cord

Cords can be attached to pendant cords or top cords or dangle end cords. They are attached to their host cord in the same way that pendants are attached to the main cord. Because these hang from cords other than the main cord, they are referred to as *subsidiary cords*. And, additional cords can be attached to them so that subsidiaries are attached to a subsidiary of a subsidiary of a subsidiary, and so on.

Combining some or all of the cord types forms a blank quipu. The number of cords on a quipu varies. There can be as few as three cords and as many as several thousand. A quipu could look like this:

Fig. 2.7

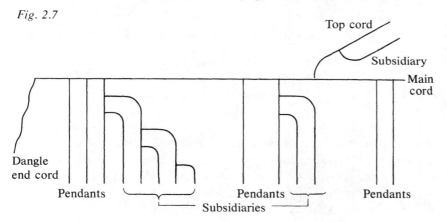

Top cord

Subsidiary

Main cord

Dangle end cord

Pendants

Pendants

Subsidiaries

Pendants

4 Several extremely important properties of quipus are inherent in the foregoing construction information. First of all, *quipus can be assigned horizontal direction.* When seeing a film, there usually are credits at one end and the word END at the other. Even if the meaning of these were not understood, they could still be used when faced with a jumble of film strips. With them, all the film strips could be oriented in the same direction. All viewing and analysis would be based on the same running direction. Therefore, terms like *before* and *after* could be applied. Similarly, a main cord has direction. Quipumakers knew which end was which; we will assume that they start at the looped ends and proceed to the knotted ends. *Quipus also can be assigned vertical direction.* Pendant cords and top cords are vertically opposite to each other with pendant cords considered to go downward and top cords upward. Terms like *above* and *below*, therefore, also became applicable. *Quipus have levels.*

Cords attached to the main cord are on one level; their subsidiaries form a second level. Subsidiaries to these subsidiaries form a third level, and so on. *Quipus are made up of cords and spaces between cords.* Cords can easily be moved until the last step in their attachment when they are fixed into position. Therefore, larger or smaller spaces between cords are an intentional part of the overall construction.

The importance of these properties is that cords can be associated with different meanings depending on their vertical direction, on their level, on their relative positions along the main cord, and, if they are subsidiaries, on their relative positions within the same level. And, just as one suspects having missed the beginning of a film when walking in on action rather than credits, quipu readers can doubt a specimen is complete if the main cord doesn't have both a looped and a knotted end, and can surmise that suspended cords are incomplete if they lack knotted tapered ends.

5 As well as having a particular placement, each cord has a color. Color is fundamental to the symbolic system of the quipu. Color coding, that is, using colors to represent something other than themselves, is a familiar idea. But color systems are used in different ways.

The colors red and green used in traffic signals have a universal meaning in Western culture. It is generally understood that red is stop and green is go. Moreover, this common understanding is incorporated into the traffic regulations of Western governments. The color system is simple and specific, and certainly no driver is free to assign his or her own meanings to these colors.

Several more elaborate color systems are used elsewhere in Western culture, for example in the electronics field. The color system for resistors, espoused by the International Electrotechnical Commission, has been adopted as standard practice in many countries. Resistors are ubiquitous in electrical equipment because the amount of electrical current in different parts of the circuitry can be regulated by their placement. In this international system, four bands of color appear on each resistor. Each of twelve colors is associated with a specific numerical value and each of the bands is associated with a particular meaning. The first two bands are read as digits (for example, violet = 7 and white = 9, so /violet/white/ = 79); the next band tells how many times to multiply by 10 (for example, red = 2, so /violet/white/red/ = 79 × 10 × 10); and the last describes the accuracy (for example, silver = 10 percent, so /violet/white/red/silver/ = 7,900 ohms plus or minus 10 percent). By combining meanings for colors with meanings for the positions, the information that can be represented has been greatly increased.

5 (a) Discussion of the resistor color system can be found on pp. 5–7 in *Reference Data for Radio Engineers,* 6th ed. (New York: Howard W. Sams and Co., 1977). This book also contains other color systems used in electronics and a discussion of the various agencies that define these systems.

(b) The law quoted is Marine Resources Law—Title 12—Chap. 17, Subchap. 4, Art. 1, Secs. 4404 and 4405 as printed in *Maine Marine Resources, Laws and Regulations,* revised to October 24, 1977, pp. 121–24.

(c) Quipu AS29, for example, uses twenty-four cord colors. Six are from different dyes and eighteen more were created by spinning combinations of these. An example of a quipu on which the color combination retains the significance of both colors is AS71 (see plate 2.3 on page 23).

Clearly, lettered signs for traffic messages and printed words on resistors would be less effective than colors. In the case of traffic messages, visibility from a distance and eliciting a rapid response are the important criteria. Locating and reading letters small enough to fit on a resistor when these components are intermingled with others in compact spaces would be difficult. Directing one's fingers to the right component is what is important, and with color coding this can be more readily done. As useful as they are, these systems are inflexible. Some group, not the individual users, defines the system and, therefore, sets its limits.

Consider another form of representation, the use of letters in physics formulas:

$$V = \frac{RT}{P}; \qquad V = IR; \qquad V = \frac{ds}{dt}.$$

In these formulas, the letter V is shorthand for volume, or voltage, or velocity because the formulas come from three different contexts within physics. Their contexts are a discussion of gases, electricity, and motion respectively. What V stands for or what each formula means, of course, depends on a knowledge of the context. We are, however, free to change the shorthand. In the first formula, which represents Boyle's Law, instead of V, T, P, R, we could use, say, a = volume, b = temperature, c = pressure, and d = universal gas constant. But, because of the behavior of gases, we are not free to change the relationship between a, b, c, d to, say,

$$a = \frac{dc}{b}.$$

So, too, a color system increases in complexity as the number of contexts it describes increase and as statements of relationship become involved.

One more example, before moving to the quipu, is the color system currently in use among lobstermen in the state of Maine in the United States. Each lobster trap that is set in the water has a colored floating buoy attached to it. By the buoy, a lobsterman can locate his traps and distinguish them from the traps of others. While the license number of the lobsterman could identify his buoy, it would probably necessitate pulling the buoy into the boat in order to read it. This would be much additional work for a lobsterman and also, if it were not his trap, could lead to misunderstanding of his intentions. The use of color enables the lobsterman to spot his buoys from afar. Also from afar, the fishing warden has to be able to observe whether a lobsterman is picking up any trap but his own. Having the colors stated on the license is inadequate for the warden's task. And so the colors must appear on the boat of the lobsterman as well as on his buoys. Color combinations have to be allowed because the number of lobstermen far exceeds the number of primary colors. Since colors can be of many different shades and combined through many different designs, legal definition or assignment of colors by the state government would limit the variety possible. And a lot of government mechanism would be needed to decide the color combinations available as different lobstermen leave or join the fleet. The solution satisfies the criteria for the system. Each lobsterman states his colors in words on the license application; he paints his buoys with those colors in whatever design pattern he chooses; and he has to mount "a buoy at least 12 inches in length with his color scheme thereon on said boat so that it will be clearly visible from each side." If a lobsterman decides that someone else who is lobstering in the same waters has colors too similar to his, one of them changes his colors. Thus, the statute continues: "Should a licensee change his buoy color design prior to obtaining a new license, he shall so notify the department of that change on a form furnished by the commissioner."

This color system relates the traps and their owner, distinguishes a lobsterman from others in his environment, and is clear but sufficiently flexible to include about five thousand individuals or more as needed. Notice that, as with the letters in Boyle's Law, the colors for a particular lobsterman can be changed but the relationship between license, trap buoys, and display buoy must be maintained. Also, as with the example of the letter V in the physics formulas, the same or similar colors can be used in different parts of the state as long as they are clear in a local context. In, for example, the resistor color system, there is no place for aesthetic expression. Lobstermen's buoys can be cheerful or dull and are even sold as decorative items to people who have no part in lobstering.

6 In the context of the traffic and resistor color systems, there is *an* answer to the question, What does red mean? But V has no fixed meaning in physics and red is associated with no specific lobsterman in Maine. However, in their local context, be it a discussion of gases or a particular port, and in association with other letters or colors, the meaning is sufficiently clear. The quipu color system, like the latter systems, is rich and flexible and of the type for which there is no one answer to such questions. Basically, the quipumaker designed each quipu using color coding to relate some cords together and to distinguish them from other cords. The number of colors on a particular quipu depends on the number of distinctions that are being made. The overall patterning of the

CODE OF THE QUIPU

colors exhibits the relationships that are being represented. The color coding of cords that are compactly connected together and likely to become intertwined, shares with the resistor color system the function of uniting the visual with the tactile. Also, recall that quipu cords can be on different levels, have different directions, and have relative positions. Another feature shared with the resistor color system is that meanings for color and meanings for positions are used in combination with each other.

Yarns dyed different colors were available to the quipumaker. Additional cord colors were created by spinning the colored yarns together. Two solid colors twisted together gives a candy cane effect, two of these twisted together using the opposite twist direction gives a mottled effect, and the two solid colors can be joined so that part of the cord is one color and the rest of it is another color. And the cord colors thus created can then be spun together creating new cord colors. With just three yarn colors, say red, yellow, and blue, and the three operations of candy striping, mottling, and joining, consider the distinctly different cord colors that are possible. There is red alone, yellow alone, blue alone; red and yellow striped, red and blue striped, yellow and blue striped; red and yellow mottled, red and blue mottled, yellow and blue mottled; red above yellow, yellow above red, red above blue, blue above red, yellow above blue, and blue above yellow. Selecting from these fifteen and using the same operations on them, there are many more.

In some cases, the quipumaker extended the subtlety of the color coding by having a two-color combination on one cord retain the significance of both colors rather than taking on a significance of its own. In these cases, a cord made of one color yarn had a small portion striped or mottled with a second color. Thus the overall cord color had one significance while the inserted color had another significance.

7 For the most part, cords had knots tied along them and the knots represented numbers. But we are certain that before knots were tied in the cords, the entire blank quipu was prepared. The overall planning and construction of the quipu was done first, including the types of cord connections, the relative placement of cords, the selection of cord colors, and even individual decorative finishings. In a few cases, quipus were found in groups mingled with other cords. Some of these quipu groups contain quipus in different stages of preparation from bundles of prepared blank quipu cords, to completely constructed blank quipus, to completely constructed quipus with some or all cords knotted. Cords with knots tied in them are only found detached when they are evidently broken.

7 Plates 2.2 and 2.3 are AS71.

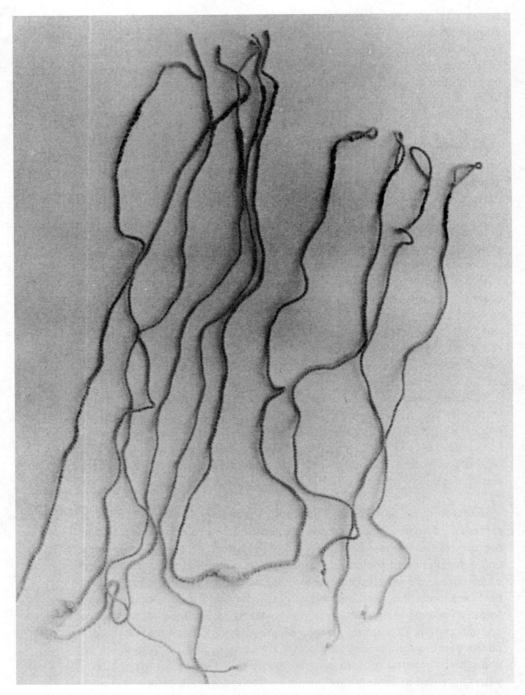

Plate 2.1. Prepared blank quipu cords. (*In the collection of P. Dauelsberg, Arica, Chile.*)

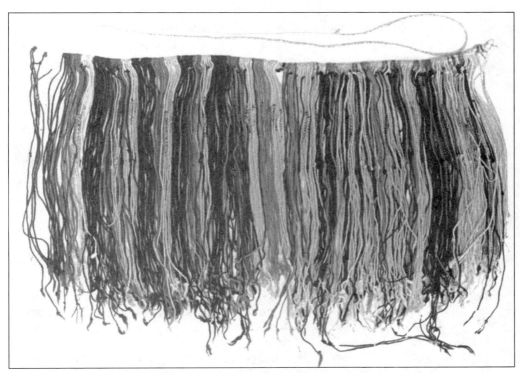

Plate 2.2. An incomplete quipu. The quipu has 251 pendants and 153 subsidiaries. Of the last 145 pendants only 1 is knotted. *(In the collection of P. Dauelsberg, Arica, Chile.)*

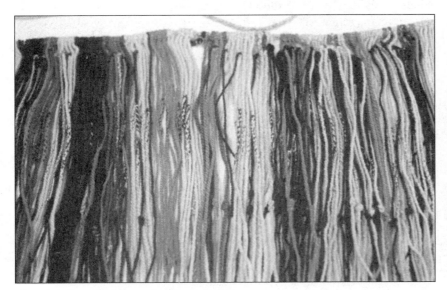

Plate 2.3. Detail of the quipu in plate 2.2.

Fig. 2.8

```
MARKET No. 8

  00.44PR:   ⋮
  00.60GR:   ⋮
  00.19PR:   ⋮
  00.99PR:   ⋮
  00.73GR:   ⋮
  00.45 TB   ⋮
       GR
  00.39PR:   ⋮
  01.97MT:   ⋮
  01.08GR:   ⋮
  00.71 TB:  ⋮
       GR
  07.55: SB  ⋮
        TL

  00.08TX:   ⋮
  07.63: SB  ⋮
        TL

  10.00:: CK
         TD
  02.37:CG   ⋮

  5974SE14
```

8 The following example is an exercise in the overall planning of a blank quipu. The problem is to design a quipu on which to record the information from the cash register tape illustrated in figure 2.8. While examining the cash register tape, the following questions are considered: What information, other than prices of individual items, is represented on the tape? Which numbers are associated with each other? Which numbers are differentiated from each other? How are the numbers associated or differentiated? What are the categories and subcategories of numbers? Note that it is not the numbers that define relationships. It is their juxtaposition, the letters following them, and their relative spacing. The tape contains different kinds of numbers: market and register identifiers, money values, and a date. Within the category of money values, there are several sorts of values: prices of purchased items (labeled PR, MT, GR, or $_{GR}^{TB}$), taxes (TX), total of prices of items and total of prices and taxes ($_{TL}^{SB}$), amount of check given ($_{TD}^{CK}$), and change (CG). Within the prices of items, distinctions are made between the types of items purchased: produce (PR), groceries (GR or $_{GR}^{TB}$), and meat (MT). The groceries are further divided into taxable groceries ($_{GR}^{TB}$) and nontaxable groceries (GR). The market identification and date also serve as beginning of tape and end of tape markers.

A quipu for this information could be laid out in a variety of ways just as different markets use different layouts for their cash register tapes. One layout is given; others could be just as valid. In our layout, different cord placement is used to distinguish between the kinds of values: individual prices, individual taxes, totals, identification numbers, and summary values. The individual prices are on pendant cords grouped together; identification numbers are on pendant cords grouped together but separated from the individual prices; the individual taxes are on subsidiary cords; the totals of individual prices and of individual taxes are with the values they sum but directed upward; summary values are on a dangle end cord. Color is used to distinguish between different types of purchased items. Since there are three types of items (groceries, produce, and meat), three different colors are used. A fourth color is used for all other cords.

A Summary Description of the Layout

1. The main cord is red (*R*). The looped end is its beginning and the knotted end is its end.

2. The first three pendant cords are red, close together, and will be for the identification information: market number, register number, date.

3. A space separates the third pendant from the next pendant.

4. The price of each item will be on a pendant cord. These cords are close together. Cords for grocery prices are yellow (*Y*), cords for produce prices are blue (*B*), and cords for meat prices are green (*G*). Each cord for taxable groceries has a subsidiary, also yellow, where the tax itself will be recorded. Centered in this group of pendants is a red top cord, for the total of the item prices. A subsidiary of the top cord, color yellow, is the total of the individual taxes.

5. Dangling from the end of the main cord, is a red cord with two red subsidiaries. The cord is for the amount of the check, the first subsidiary for the total cost, and the second for the change.

Look at figure 2.9 for a schematic of the layout.

Fig. 2.9. A schematic of the layout.

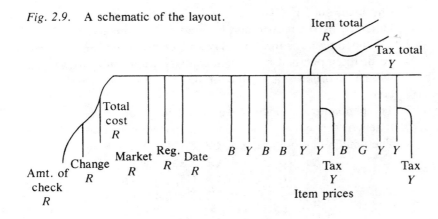

Exercise 2.1
Design a quipu on which to record the information from figure 2.10.

Fig. 2.10

Auto Service Hours for the Next 4 Weeks						
Mon.	Tues.	Wed.	Thur.	Fri.	Sat.	Sun.
9-5	9-5	9-7	9-9	9-9	9-5	1-5
9-5	9-5	9-7	9-9	9-9	9-9	Closed
Closed	Closed	9-7	9-9	9-9	9-5	1-5
9-5	9-5	8-1	8-1	9-9	9-5	1-5

An Answer to Exercise 2.1

Since for each day there are two values, an opening hour and a closing hour, a pendant cord with a subsidiary can represent each day. Spacing can be used to associate the pendant cords for one week and distinguish them from those for the next week. A day the station is closed needs special consideration. If a pendant cord and subsidiary are included for the day but with no numbers on them, there could be ambiguity. It might be interpreted to mean that the hours are not yet decided or that the quipu is not yet complete. What if the cords are simply omitted? If the cords are color coded such that each day is represented by a different color, fewer than seven pendants in a group will present no problem. Without the color identification, it would not be clear which day was being omitted. So seven different colors are needed. Call them C1, C2, C3, C4, C5, C6, C7.

1. The main cord is C1 and, as usual, its beginning and end are indicated by the looped and knotted ends respectively.
2. The opening hour for each day will be recorded on a pendant cord. Each pendant has one subsidiary where the closing hour will be recorded. Each pendant and its subsidiary are the same color.
3. The pendants are placed seven close together, space; six close together, space; five close together, space; seven close together.
4. The pendant colors are: C1, C2, C3, C4, C5, C6, C7;
   C1, C2, C3, C4, C5, C6;
   C3, C4, C5, C6, C7;
   C1, C2, C3, C4, C5, C6, C7.

9 If, in answer to, How many sheep are there? one pebble is placed on a heap or one chalk mark is made as each sheep passes, the respondent is carrying out a matching process. Each individual object—a pebble or a chalk mark—is matched with a sheep. The knots on a quipu are not of this type: they are not a collection of knots, each knot matched to an individual object. Rather, the knots form a symbolic representation. The use of symbolic representation indicates that numbers have become abstracted from their association with particular objects.

The symbolic representation of numbers on quipus is of an exceptionally sophisticated type. In fact, although the symbols are made up of knots rather than written figures, it is basically the same as ours. It is a base 10 positional system. Our ease in translating numbers on quipus is a tribute to the fact that our early arithmetic education resulted in our incorporation of the rules of a base 10 positional system. When using our own system, most of us do not recognize that we are applying rules.

9 The words *sign* and *symbol* are used by us in the following way. We use sign for a figure, mark, character, or object, and symbol when that figure, mark, character, or object stands for something by definition, association, or convention among some group of people. As an example, let us take an ampersand (&). The ampersand is a sign. Conventionally, it stands for the word *and*. If we call it a sign, we mean solely the character. We call it a symbol when we also acknowledge that it has a conventional meaning. We know when seeing "John & Bill" not to verbalize it as "John ampersand Bill" but as "John and Bill." And, when & stands out of context, we could distinguish the sign from the symbol by using for the former its name (ampersand) and for the latter the word (and). Math-

We also tend to overlook the fact that our system is but one of many possible systems, and that its development took about four thousand years with major contributions from the Babylonians, Hindus, and Arabs.

To clarify what is meant by our base 10 positional system, consider the meaning of *1064*. In our system, only 10 individual symbols are used: 0, 1, 2, 3, 4, 5, 6, 7, 8, 9. To express any value, we select from these and place them in some order. The value of the collection depends on where the individual symbols are placed. The collection 1064 has the value $(1 \times 1,000) + (0 \times 100) + (6 \times 10) + (4 \times 1)$. Each consecutive position, moving to the left, is multiplied by 10 another time. The number of times 10 is multiplied is called the *power* of 10, and so each consecutive position is one higher power of 10. The concept underlying a base positional system is not tied to the base 10. Any positive integer can be used as a base. For example, if the base is 6, only 6 individual symbols are needed. The value of any collection of these depends on where each symbol is placed. In base 6, each consecutive position, moving to the left, is a higher power of 6. Hence, using this positional rule and our symbols 0, 1, 2, 3, 4, 5, any value can be represented. The collection 4532 interpreted in the base 6 positional system means $(4 \times 216) + (5 \times 36) + (3 \times 6) + (2 \times 1)$. Notice that it has the same value as that which was written in base 10 as 1064.

10 To appreciate the significance of a base positional system, consider the system of roman numerals which is not of this type. It was one of the systems in use in Europe through the seventeenth century and is still used in some contexts in Western culture. The individual symbols are I, V, X, L, C, D, and M. In our notation, the values of these are 1, 5, 10, 50, 100, 500, and 1,000 respectively. The value of a collection of these symbols depends on where they are placed. However, the evaluation depends, not on the position of individual symbols per se, but on the relationship of the value of a symbol to the value of nearby symbols. When symbols have decreasing value moving to the right, the value of the collection is their sum. For example, XVI translates into our 16 because it is interpreted as $10 + 5 + 1$. When a symbol of smaller value precedes one of larger value, it is subtracted. Hence, IX translates into our 9. In the collection MCMXLIV, both the additive and subtractive principles are involved. A direct translation of each symbol in this collection is 1,000; 100; 1,000; 10; 50; 1; 5. Since 100 precedes a 1,000 and is less than it, and 10 precedes 50, and 1 precedes 5, the collection is translated as $1,000 + (1,000 - 100) + (50 - 10) + (5 - 1) = 1,944$. The formation of a symbol collection is less clear; there are no exhaustive explicit rules. For example, the value written in our system as 80 could be written in roman numerals as XXC (interpreted as $100 - 10 - 10$)

ematical symbols are arbitrarily chosen signs that have a defined or conventionally accepted meaning to some group. They need some further clarification. The marks or squiggles 3 and 5 are signs. However, once values are associated with them and they are used to stand for those values, they are symbols. We know when seeing "Subtract 3 from 5" that it is the values that are involved. But the sign 3 (solely the character) and the symbol 3 (the character with an associated value) have the same verbalization. So, we call it the symbol 3 when we mean to evoke the value, and when we refer, for example, to the symbol X used by the Romans, we mean to imply that it had a value according to a group of users.

9–11 For further discussion of number systems see:
(*a*) Florian Cajori, *A History of Mathematical Notations,* vol. 1 (1928; reprint ed., LaSalle, Ill.: Open Court Publishing Co., 1974).
(*b*) Tobias Dantzig, *Number: The Language of Science,* 4th ed. (New York: The Free Press, 1967). (Our citations are from pp. 24–25, 29–30.)
(*c*) A. Seidenberg, "The Diffusion of Counting Practices," *University of California Publications in Mathematics* 3 (1960): 215–300.
(*d*) D. E. Smith and J. Ginsburg, "From Numbers to Numerals and from Numerals to Computation," in *The World of Mathematics,* ed. James R. Newman, vol. 1 (New York: Simon and Schuster, 1956); pp. 442–64.
(*e*) Raymond L. Wilder, *Evolution of Mathematical Concepts* (New York: John Wiley and Sons, 1968).
For extensive examples of number systems not discussed in the above references, see W. C. Eells, "Number Systems of the North American Indians," *American Mathematical Monthly* 20 (1913), pt. 1, pp. 263–72; pt. 2, pp. 293–99. For a specific discussion of zero, see Carl B. Boyer, "Zero: The Symbol, the Concept, the Number," *National Mathematics Magazine* 18 (1944): 323–30.

or as LXXX (interpreted as 50 + 10 + 10 + 10). LXXX is used because, in general, only a single symbol of lesser value immediately precedes a symbol of greater value.

An important contrast between a base positional system and the roman numerals, is that in a base positional system, no new signs need be introduced as the magnitude of the numbers grow. With the roman numerals already given, it is, to say the least, inconvenient to write a collection that translates into our number 120,000. A collection of 120 M's would be needed. Although they are rarely seen now, different signs were used to indicate thousandfold. One of them was a horizontal stroke above a number. Thus, since CXX is equivalent to our 120, $\overline{\text{CXX}}$ would be equivalent to our 120,000. Notice also, that the size of a roman numeral representation is not related to the magnitude of the number. For example, the single sign M is greater in value than the collection XXXVIII. In a base positional system, the size of the representation and the value are always related: 1,000 is greater in value than 38.

Most importantly, a base positional system simplifies arithmetic. Because there are a limited and specific set of symbols and explicit rules for forming them into other symbols, general arithmetic principles can be developed and explicitly stated. The fact that quipus use a base positional system for numbers is evidence that the Incas knew the key concepts involved in arithmetic.

In the authoritative work *Number: The Language of Science,* when discussing the history of number systems in Western culture, Danzig emphasizes the relationship between a system of number representation and arithmetic. He first asks: Which system of numeration, the Greek or the Roman was the superior? But he immediately dismisses the question because it misses the point. His pointed question is: ". . . how well is the system adapted to arithmetical operations, and what ease does it lend to calculations?" Neither of their systems has advantage because ". . . neither was capable of creating an arithmetic which could be used by a man of average intelligence." "That is why," he continues, "from the beginning of history until the advent of our modern *positional* numeration, so little progress was made in the art of reckoning." And Danzig concludes: "When viewed in this light, . . . the *principle of position* assumes the proportion of a world-event. Not only did this principle constitute a radical departure in method, but we know now that without it no progress in arithmetic was possible."

In a base positional system, the number chosen for the base is arbitrary. Often the choice depends on the particular context in which the system is used. For example, most modern digital computers work in base 2 (binary). With base 2, only 2 different signs are needed and each consecutive position in a collection is evaluated as one higher power of 2. Two signs can easily be distinguished electronically: one can be "current on" and the other "current off," or current flows in opposite directions can be used. All internal machine arithmetic is then manipulation of these current states. Many people who work directly with computers become adept at the use of base 8 (octal) because it can rapidly be converted into base 2 but the representation of a number is much more concise. And some people believed so strongly in the scientific superiority of a base 12 system that an organization, the Duodecimal Society of America, was formed devoted to its propagation.

From the preceding discussion, it should be clear that although the base 10 positional system is most familiar to us, the Inca use of it was neither necessary or to have been expected. They might not have developed a positional system at all, and even with a positional system, they need not have chosen the base 10.

11 On each cord there are clusters of knots. The collection of clusters on each cord form a symbolic representation of a number. Each cluster contains 0 to 9 knots and the clusters are separated by spaces that distinguish one cluster position from the next. Each consecutive cluster position, moving from the free end of a cord to where it is attached to another cord, is one higher power of 10. Moreover, the value of a particular cluster position is further clarified by the type of knots used. Long knots (L) are used in the units position and single knots (s) are used in all other positions. Since a long knot cannot be made with fewer than 2 turns, a 1 in the units position is represented by a figure eight knot (E). These knots are formed as shown in figure 2.11. A pendant

11 Locke's conclusion is reported in L. L. Locke, *The Ancient Quipu or Peruvian Knot Record* (New York: American Museum of Natural History, 1923).

Fig. 2.11

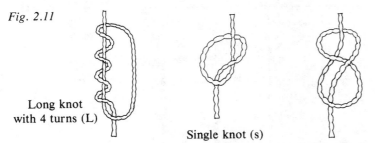

Long knot with 4 turns (L)

Single knot (s)

Figure eight knot (E)

cord with three cluster positions containing 4 single knots, 5 single knots, and a long knot of 2 turns respectively when read downward (see fig. 2.12) would be interpreted and written in our notation as

$$452 = (4 \times 100) + (5 \times 10) + (2 \times 1).$$

Fig. 2.12

Crucial to a base positional system is a representation for "zero." Clearly, our number 407 is different in value from our number 47: a sign for none or nothing is placed in the second position in order to have the 4 fall in the third position. The concept of zero can be divided into three parts: first, the understanding that positions containing nothing contribute to the overall value of a number; second, that there must be a way of representing nothing; and third, that when the representation of nothing stands by itself, it is also a number. On quipus, zero is represented by having no knots in a cluster position. The more carefully the cluster positions are aligned from cord to cord, the more apparent is an empty position on one cord when related to the others. Our numbers 370; 0; 2,164; and 601 are represented on pendant cords as diagrammed in figure 2.13.

Fig. 2.13

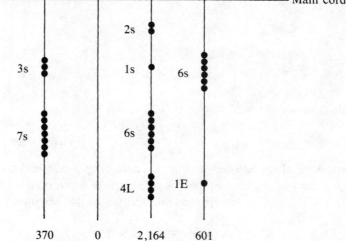

CODE OF THE QUIPU

Since the highest valued position is always closest to the cord connection, knot clusters on subsidiaries are not necessarily aligned with the clusters on pendant cords. For the same reason, the values of knot clusters on top cords are read in the direction opposite to pendant cords. Examples are shown in figure 2.14.

Fig. 2.14

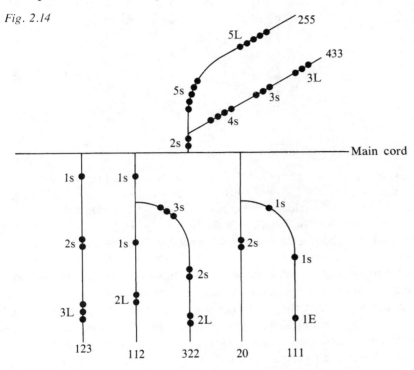

The fact that numbers are represented with a base 10 positional system was established by Leland L. Locke at the beginning of this century. He noted that, if knots are interpreted in this way, when top cords are present on a quipu, the numbers on the top cords are usually the sum of the numbers on the pendant cords with which they are associated. This relationship confirmed the interpretation.

12. On several quipus there is a somewhat different way of placing numbers on the cords. Nevertheless, the base 10 system is still maintained. These quipus have two or three numbers on a single cord. Because of the alignment of the knot clusters and because the units digit is distinguished by a long knot or figure eight knot, it remains clear where one level of numbers end and the next begin. An example of this form of number placement is shown in figure 2.15.

12 A few examples of quipus with several numbers per cord are AS4, AS80, AS91, AS173, and AS174. This interpretation is verified by their arithmetic relationship to numbers on other cords.

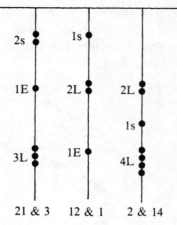

Fig. 2.15

Main cord

21 & 3   12 & 1   2 & 14

Exercise 2.1 (Auto Service Hours) is an example in which more than one number per cord would be useful. In the exercise, there was an opening hour and a closing hour for each day. An alternate solution to the exercise would be to place both on the same cord with the higher number being the opening hour and the lower one the closing hour.

13 Numbers are fundamental to the symbolic system of the quipus. Numbers, however, can express different things. They frequently express quantities, but they can also serve as labels. A common way of labeling objects with numbers is to place a collection of objects in some specific order and then assign consecutive integers to them. In this way, number labels are given to houses on a street, products made by a single company, people applying for a service, or items in a catalog. For example, on the cash register tape illustrated in figure 2.8, the market is identified as "Market No. 8." Number labeling is carried further when a number label is a composite of several identifying numbers. An apartment in a building may be labeled 412 when it is the twelfth apartment on the fourth floor. The building may have only ninety apartments, that is sixteen on each of five floors. Thus, a composite number label uses the symbols of our base 10 system and the placement of the digits is important, but the number as a whole has no quantitative interpretation.

14 Thus, depending on what is to be recorded, the assigning of numbers as labels and the assembling of numbers that express magnitudes are part of the planning of a quipu. The sizes of the numbers have to be considered in order to select the length of the cords to be used. To tie as many as nine knots in each of four of five knot clusters and have space between the clusters, requires a longer blank cord. And, the max-

14 Plates 2.4 and 2.5 are AS34. Plate 2.6 is AS38.

imum number of clusters on any cord of a particular quipu determines the placement of the clusters on the other cords of the quipu (see, for example fig. 2.13). After deciding the placement of the knot clusters, the knots must finally be tied into the cords of the prepared blank quipus. From the partially completed quipus, it is evident that the cords were not necessarily knotted serially from one end of the quipu to the other.

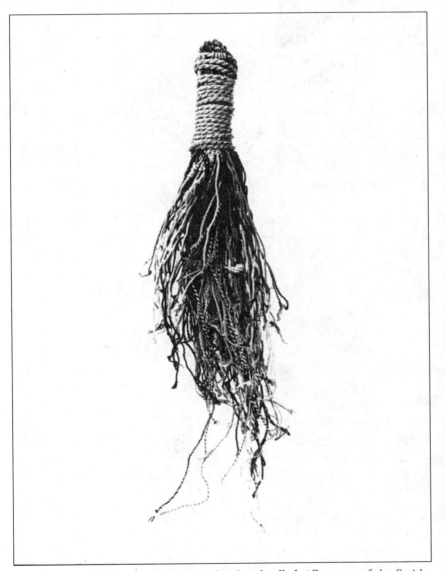

Plate 2.4. A quipu that has been completed and rolled. *(Courtesy of the Smithsonian National Museum, Washington, D.C.)*

Plate 2.5. The quipu of plate 2.4 unrolled. *(In the collection of the Smithsonian National Museum, Washington, D.C.)*

Plate 2.6. A completed quipu. Observe the details of cord and knot construction and placement. *(In the collection of the Museo National de Anthropología y Arqueología, Lima, Peru.)*

CODE OF THE QUIPU

15 We can now complete the market tape example by including, on the schematic (fig. 2.16), the placement of knot clusters and the number and type of knots in each cluster. (There is no need to draw each knot. We use one mark for the position of each cluster.)

Fig. 2.16. Schematic for figure 2.8.

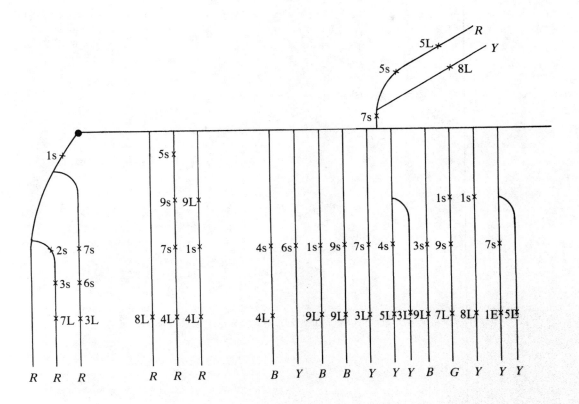

Chapter 3 | Inca Insistence

1 Walk into any museum devoted to the study of past cultures. There, on display, you will see collections of things. Taken together, these things document global achievement; they can be seen as alternative solutions to panhuman problems such as crossing water, coming to terms with the unknowable, and keeping oneself warm and protected.

In some ways, the boats, prayer shawls, and clothing in a museum room devoted to one culture will be similar to objects intended for the same purpose found in another room given over to a different culture. A boat is intended to stay afloat. To do so, it must be built within the constraints of physical law. Likewise, the biology of the human body puts limits on certain aspects of clothing; for example, a shirt must have a hole through which a person's neck and head can protrude. Because similarly strong constraints are not involved for religious objects, more variety is found as one moves from room to room, from culture to culture. Another way to account for similarity across cultures is history. Often enough, a particular kind of boat, or shirt, or prayer shawl, or just the idea of how to make one of these things, was brought from the people of one culture to the people of a different culture.

Within any one room, the artifacts of a particular culture will be distinctive. Part of the distinction is due to factors such as climate; clearly, one needs bulkier clothing to live in the arctic than one needs in a tropical forest. Yet there is something more to it. For example, the objects in the display cases of one culture may tend toward the airy and light; elsewhere, in another room given over to a different culture, similar objects may tend toward the closed and massive. More often, the distinction is subtle; intuitively one senses being close to a particular way of doing things that, however difficult to define, seems to be present, strong, and quite real.

As one looks more closely at the things produced in a culture, an object may take on special significance. We may not know the process of thought that leads to this partially conscious settling on an object. It may even be difficult to state why one object, rather than another, becomes so important in our mind, or more precisely, in our imagery. But this experience is common, and often different people can agree that this thing, and not that one, can express the rest. Imagine a very large room filled with the things produced in our culture. The room contains advertisements, oil, and a bar of steel. In order to accommodate large things, there are scale models of shopping malls, billboards, and bridges. A visitor to this room—a person from a very different culture—is at first confused by the thousands of items. After several visits, he comes away with the image of an automobile. If asked to explain, he might

1 For a further discussion of artifacts that take on peculiar significance, see Robert Ascher, "Tin Can Archaeology," *Historical Archaeology* 8 (1974): 7–15.

answer that the automobile stands for, sums up, or expresses his museum experience of the American branch of Western culture in the last quarter of the twentieth century. Many of us can comprehend his answer and perhaps agree that his selection has merit. For the outsider, such objects are a key to the culture; for the insider, they are a door into larger areas of meaning. They can be common and replaceable like the automobile, fixed in place like the medieval cathedral, or impermanent and leave almost no trace, like the igloo. Such objects often sum up a culture in a way that words cannot.

To sense that the objects of one culture are airy and light, while in another, similar objects are dense and massive, is a step toward the idea of insistence. The important thing about a culture is its particular insistence; in fact, that is what defines a culture. The quipu is an interesting and important object, worthy of attention in itself. But it also serves to express Inca insistence in one compelling and concrete image. We shall need to amplify the general idea of insistence. Then, what characterizes Inca insistence will be explored. How the quipu expresses Inca insistence will be understood at the end of the exposition.

2 In 1934–35, Gertrude Stein delivered a series of lectures in the United States. Her notion of insistence specifically appears in the lecture entitled "Portraits and Repetition," *Lectures in America,* (New York: Random House, 1975), pp. 163–206. But how she employs insistence is found in all lectures published in that book.

2 Gertrude Stein, the American novelist and essayist, is best known for her comment on a rose. A rose is a rose is a rose, she said. Throughout her prose, Stein used thousands of such constructions. Critics who took her to task thought that her writing was repetitious. In a talk given in 1934, Stein responded to the critics. What they took to be repetition was not that, but rather it was something she named insistence. At one place in her talk, she illustrated her point with the case of a frog. A frog hops many times but the hops are not repetitious: the hops are the frog's way of insisting. Later in her talk, she recited several previously published word portraits or sketches of individuals. What makes one person different from another, according to Gertrude Stein, is their individual insistence, so it is insistence that she captured in her word portraits. Then she went on to use the term in connection with civilization: ". . . when you first realize the history of various civilizations, that too makes one realize repetition at the same time the difference of insistence. Every civilization insisted in its own way before it went away." We rephrase Stein's difficult prose this way: civilizations share much in common and hence they seem to repeat each other, but each civilization "insisted in its own way."

When struck by insistence, say that of an individual, it is often hard to specify the source of the affect. Trying the analytic approach does not solve the problem. Break up what has been perceived into its constituent parts, examine each part independently, verbalize the results,

CODE OF THE QUIPU

and still the problem is not solved. Is it the way she walks, or moves her head, or how she uses her hands while making a point? No enumeration of features, however lengthy, can resolve what was rightly perceived as a whole. Mere addition gets no further. Surely the movements of her feet, hands, and head, go into the making of her particular insistence, but taken independently or in sum, they are not it per se. However, we can say that her movements are some of the things that are characteristic of her insistence.

In the case of a person, as in a culture, insistence often shows up in details. Two friends recognize each other from small cues that others miss. A fleeting gesture is sufficient for them, while others, less familiar with their particular insistences, need to know more. A person who is an expert in a different culture can often recognize that culture in a fragment of cloth. Occurrences like these suggest some generalizations. First of all, insistence is pervasive; it appears in the small as well as in the large. Second, the facility to identify it increases with knowledge of the whole, whether it be a person, a society, or a culture. Furthermore, every time it is successfully identified from a detail, the concept itself passes a test. There is nothing mysterious in the expert identifying a culture from a bit of cloth: he may not be able to spell out the processes that led to the identification, but no matter, it worked. The fact that it worked, and it works in a multitude of familiar daily instances, tells us that insistence truly exists in the world.

The expert in the ways of a culture other than his own cannot always identify the culture from a fragmentary piece. It may be his failure, but more likely the distinctive insistence is weak or simply not in the fragment. Care must be taken to steer clear of the notion that insistence is monolithic; actually its strength varies from nothing to very forceful. In every culture there are alternative ways of doing things; there is always exception, variety, and idiosyncracy, and insistence behaves accordingly. There is also bound to be change through time, and as culture changes, so does the patterning of insistence. Differences in strength, then, vary with what is observed, and when the thing observed materialized in the history of the culture.

Just one more general notion needs to be amplified. Newspapers devote separate pages or even entire sections to politics, cooking, sports, business, fashion, and so on. In spite of the separation, insistence may surface under any number of these headings. That is to say, it does not obey the boundaries of the categories into which a culture is ordinarily classified. In the example given earlier, the insistence of a woman was picked up from the movements of her feet, hands, and head. For these, substitute the way she wears a hat, laughs, and writes a note. Or mix

3 The fresco shown in plate 3.1 is *The Annunciation: Submission,* by Fra Angelico. Our discussion of how a fifteenth-century Italian may have viewed a painting draws upon the work of Michael Baxandall, *Painting and Experience in 15th Century Italy* (New York: Oxford University Press, 1972). The relative merits of artifacts and words is considered by Craig Gilborn in "Words and Machines: The Denial of Experience," *Museum News,* September 1968, pp. 25–29. A reconstruction of the aftermath of the Spanish conquest from the point of view of the conquered is found in Nathan Wachtel, *The Vision of The Vanquished: The Spanish Conquest of Peru Through Indian Eyes* (New York: Harper and Row, 1976).

them, taking for example, how she laughs, walks, and wears a hat, and the same holds true. In the case of a culture, insistence cuts across any arbitrary category and in so doing provides a measure of cultural coherence.

3 Thus far, the discussion has drawn largely upon an intuitive understanding of insistence. Now we turn to the problem of finding the particular insistence of a culture that no longer exists. Unlike insistence in the present which is accessible through immersion in a living culture, past insistence is perceived only through visual and tactile evidence that is lifeless and disjointed. To find what is characteristic of Inca insistence is especially difficult, as illustrated by the following comparative example.

The example is drawn from Italy in the fifteenth century. In that century, the Incas flourished in another part of the world, so the same amount of time separates us from both cultures. The Incas survived for about one hundred years, and so did the Italian Renaissance; hence the cases are comparable in that respect also.

The example is a fresco painting (plate 3.1). You may recognize the subject matter of the painting. If so, it is because it arises out of a religion that persists around you today. Nevertheless, the understandings and experiences that a contemporary man brings to the painting are very different from those brought by a fifteenth-century Italian. He saw many paintings on the same subject, and he saw them inside churches, with their special aura, not in museums or books. Having been prepared by listening to sermons, he saw that the painting not only depicts the meeting of Gabriel and Mary, but also shows a particular well-defined stage in the spiritual changes in Mary as she spoke with the Angel Gabriel. We know this because sermons as well as pictures survived, and they were intended to complement each other. Acquaintance with Christian beliefs, such as the Annunciation, were further reinforced for the fifteenth-century Italian by hearing and watching religious processions, receiving the communion wafer, and by observing the ever present clerics in the marketplace. The colors of the fresco are now faded. We account for the change by the passage of time. The fifteenth-century person, seeing a faded painting, might pass a remark on the wealth of the person who paid for it. A picture was not created and then purchased; rather, a buyer would specify, in advance, what a picture should be about, perhaps how large it should be, and other details, including the quality of the paint. This is known from surviving commercial contracts. Many a buyer tried to insure in writing that the colors in his picture, especially the blue, did not fade or flake soon after the picture was completed. But sermons and contracts can take us just so far: look at how the Angel Gabriel and Mary hold their hands. There is no way to know how the fifteenth-century Italian interpreted these gestures.

Now take the same painting and change the circumstances. There is no writing as we understand it, set down by a person who was a member of the culture. Commentary provided by the written sermons and contracts are lacking. The religion ends abrubtly not long after the close of the fifteenth century. However, there are some writings that were composed by the same people who were parties to the destruction of the religion and the culture. Other kinds of nonverbal evidence, for example, deserted buildings, are plentiful. And now, let us look at the painting again and attempt to tell what it means, keeping the revised circumstances in mind.

Plate 3.1. The Annunciation: Submission, by Fra Angelico. *(Courtesy of Fratelli Alinari, Florence, Italy.)*

Three people are seen. Two of them are more or less facing each other, so they may be having a conversation. Perhaps the kneeling person is deferring to the one who is standing. The man at the far left seems to be watching. Large circles are drawn around the heads of the three people, but we do not know why they are there. One of the individuals has the wings of a large bird protruding from her (?) back; these are harder to account for than the circles. We could go on, but it is already clear that under the revised circumstances, we cannot really tell what the picture is about.

The revised circumstances describe an actual situation, namely what happened to Inca religion and culture. Lacking knowledge of Inca religion, were it an Inca painting, the contemporary viewer could bring little that would help him to know its meaning. Like the fifteenth-century Italian, an Inca brought to the picture his own culture's predispositions, his experiences, and knowledge of the myths and symbols of his religion. But unlike an Italian picture, an Inca picture cannot be surrounded by illuminating commentary. Put directly, the same number of years stand between us and both the Italian Renaissance and the Inca state, but the absence of native commentary in the latter case makes it impossible to achieve anything close to the same kind of reconstruction. In their writings, the Spanish conquerers of the Incas necessarily present the views of the outsider. If the outsiders were merely interested observers, we would be inclined to accept more of what they wrote. Few Spaniards were that, Cieza de León being one of the exceptions. Coming many years after the Inca demise, there were writers whose inheritance was partially Andean. Their environment and education were undeniably colonial Spanish. In a later chapter we will have occasion to call upon the pictoral testimony of one such person.

So artifacts are the only testimonials by Incas to themselves. Using artifacts as a point of departure can take us far in the direction of finding Inca insistence. Recall that insistence is present in artifacts as well as in words. Artifactual evidence is filled with gaps; the idea of insistence is useful here because it crisscrosses categories of culture, is tolerant of incompleteness, and is sensitive to significant detail. Insistence changes slowly in a traditional culture. The Inca state lasted for about a century. Over that short a period, major shifts are unlikely. The insistence of a culture gives it definition; thus, while allowing for exceptions and variety, a coherence that was really there can emerge from disparate evidence. While emphasis is on Inca artifacts, Spanish writings are not ignored. Especially in those areas where artifactual evidence overlaps Spanish writings, we do not hesitate to draw upon them.

4 Cuzco was the single largest artifact of the Inca state. It is situated in an eleven-thousand-foot-high valley on the eastern slope of the Andes not far from the equator. The Incas had lived in and around Cuzco for about two hundred years before beginning their outward expansion. During this early period, Cuzco might better be referred to as a collection of artifacts: houses, shrines, and streets. But a unifying plan was imposed upon the site soon after the Incas started their conquests. Certain factors called for adjustments in their ideal plan. The valley is elongated and the terrain is irregular. Some of the man-made structures that would be retained could not be moved. In any case, the Cuzco that resulted from the plan can be seen as a deliberate expression of Inca organization in spatial terms. Concern with *spatial relations* is characteristic of Inca insistence.

The meaning of a few native terms is prerequisite to an understanding of spatial relations in Cuzco. *Huaca*, the first term, can be translated as an object that is sacred. Sacred objects were numerous and varied; some examples are caves, special piles of stones, tombs, mountain peaks, battle fields, springs, particular buildings, and quaries. Huacas were not peculiar to the Incas. The notion of sacred objects in the same sense occurs throughout the Andes and predates the Inca state. The second term is *ayllu,* and like the first, the idea was widespread and older than the Inca state. It was applied to human relationships. For example, several families claiming a common ancestor formed an ayllu. The members of such an ayllu lived in the same territory. In an agricultural region, they worked and owned the land communally. Used in this sense, the ayllu was the fundamental social unit beyond the family. A village, a town, or a city might be composed of several ayllus. In that case, they were grouped into two units, half the ayllus in one, half in the other. Each of the ayllus retained its separate identity within the larger units.

With this terminology, we return to the spatial arrangement of Cuzco. For clarity, a diagram is constructed as we progress through the arrangement. Cuzco was divided into four quarters.

Quarter   Quarter   Quarter   Quarter

Each quarter would eventually become the abode of three ayllus.

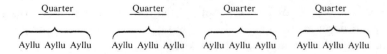

Quarter   Quarter   Quarter   Quarter

Ayllu Ayllu Ayllu   Ayllu Ayllu Ayllu   Ayllu Ayllu Ayllu   Ayllu Ayllu Ayllu

4 The ancient city of Cuzco is described by Manuel Chávez Ballón in "Cuzco, Capital del Imperio," *Wayka*, no. 3 (University Nacional del Cuzco, Cuzco, Peru, 1971): 1–14; and John H. Rowe, "What Kind of a Settlement was Inca Cuzco," *Ñawa Pacha* 5 (1967): 59–76. An interpretation of the meaning of the conceptual lines and their names is proposed by R. T. Zuidema, *The Ceque System of Cuzco: The Social Organization of the Capital of the Inca*, International Archives of Ethnography (Leiden: E. J. Brill, 1964).

In Cuzco, the common ancestor claimed by an ayllu was a Sapa Inca, or a head of state. He had one principal wife and several secondary wives, so an ayllu could become quite large in a short time. The ayllus were grouped into two larger units, called Upper Cuzco and Lower Cuzco.

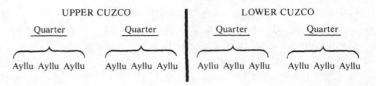

Upper Cuzco was located in the northern area of the city and Lower Cuzco was in the south. In and around Cuzco, there were approximately four hundred huacas. These were connected by conceptual lines radiating from the center of the city. Each of the ayllus in three of the four quarters was charged with the care of the huacas along three lines, while each of those in the fourth quarter had the care of the huacas along five lines. Thus, with each of three ayllus in three quarters in charge of the huacas along three lines, and each of the three ayllus in one quarter in charge of huacas along five lines, there were forty-two lines. In effect, these conceptual lines connecting real objects divided the city into sacred spaces.

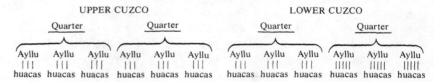

5 The Incas called their state Tawantinsuyu. This is translated Land of the Four Quarters. The term must be taken literally, for the Incas indeed divided the lands under their control into four parts, just as they had partitioned Cuzco. And from the central plaza in Cuzco, there emanated the roads that connected the capital with all parts of the four quarters. The Incas, as all conquest states, were interested in moving troops, and the roads served that and other ends. Inca roads were remarkable in number, construction, and length, but they do not, in themselves, lead to an understanding of what was particularly Inca.

Portability, rather than mobility—although they surely had that too— was characteristic of Inca insistence. Things that moved along the road weighed no more than could be carried by one person. An Inca walked with his burden on his back. The standard artifact used for carrying was

5 A brief discussion of Inca furniture appears in Wendell C. Bennett, "Household Furniture," *Handbook of South American Indians,* ed. J. H. Steward, vol. 5, Smithsonian Institution, Bureau of American Ethnology Bulletin no. 143 (Washington, D.C., 1949), pp. 21–27; and an early treatment of the aryballos is Marshall H. Saville, *The Pottery Arybal of the Incas,* Museum of the American Indian, Heye Foundation, Notes and Monographs, no. 3 (New York, 1926), pp. 111–19. For the vessel in an archaeological context, see Luis A. Llanos, "Trabajos Arequeologicos en el Dep. del Cuzco," *Revista del Museo Nacional* 5 (Peru, 1936): 123–56.

a piece of strong cloth. The cloth was placed on the ground and the load put in it. Two opposite corners were folded over the load. The other two corners were grasped, the load was lifted onto the back, and these corners were tied at the chest. So loaded, Inca men and women moved along roads that were often steeply inclined. The only exceptions were members of the Inca elite: they were carried in litters by walking men. The distance between way stations along an Inca road varied from twelve miles in the mountains to eighteen miles on level ground. These distances reflect the need for frequent rest and resupply. It is true that one animal—the llama—carried burdens. But llamas cannot be loaded with more weight than a man can carry, people did not ride on their backs, and they must stop frequently.

Nowhere was portability more marked than in the interior of every house. Regardless of class, rank, or other distinctions, houses were essentially devoid of furniture. In some instances, a bench made of stone or adobe was fixed along an interior wall. People squatted or stood while awake. They went to sleep on the floor, in their clothing, wrapped in a blanket. The kitchen ware included a variety of plates, pots, jars, and laddles, all of them easily transportable. Significantly, the most distinctive Inca vessel was designed for travel. The aryballos—as it is called after the Greek vessel to which it bears a superficial resemblance—is flared at the rim, narrow at the neck, wide in the body, and tapered sharply at the bottom. A knob projects from the top of the body close to where the neck begins. There are two vertical handles on opposite sides of the body where the taper starts. A rope was passed through one handle, then up and over the knob, and back down and through the second handle. With the rope in position, the vessel was portable (plate 3.2). It was placed on the back, the loose ends were tied at the chest, and thus fixed in an upright position on the back of the bearer. In short, an Inca family could walk away from their house taking all their possessions with them. Many families did that. The Inca authorities moved people, en masse, from a region under their control to a newly acquired one. The presence of the colonists in the new area presumably upset pre-Inca political units, and at the same time, introduced Inca ways.

6 *Cloth* was so important in Inca culture that it was characteristic of Inca insistence in and of itself. At every turning point in the life history of an individual, cloth played a key role. A gift of fabrics and food was appropriate for a child when he or she was weaned; a male put on a breechcloth for the first time when it was ceremoniously given to him at the point in his life that he turned from boyhood to manhood; a dead body was wrapped and buried in multiple layers of textiles, and for

Plate 3.2. Sculpture of man with aryballos in position. *(Courtesy of the Museum für Völkerkunde, Berlin, Germany.)*

6 The paramount importance of cloth is pointed out in John Murra, "Cloth and Its Functions in the Inca State," *American Anthropologist* 64 (1962): 710–27. Another important discussion of textiles is A. H. Gayton, *The Cultural Significance of Peruvian Textiles: Production, Function, Aesthetics,* Kroeber Anthropological Society Papers, no. 25 (Berkeley, 1961), pp. 111–28. Earlier basic works on textiles are found in several sources; the most useful are: M. D. C. Crawford, *Peruvian Textiles,* American Museum of Natural History, Anthropological Papers, vol. 12 (New York, 1915), pp. 52–104; Lila O'Neale and A. L. Kroeber, *Textile Periods of Ancient Peru,* University of California Publications in American Archaeology and Ethnology, vol. 28 (Berkeley, 1930); Lila O'Neale and B. J. Clark, *Textile Periods of Ancient Peru II,* ibid, vol. 40 (Berkeley, 1948); and P. A. Means, *A Study of Peruvian Textiles* (Boston: Museum of Fine Arts, 1932). A study largely dependent on the work of John Murra is Barbara Braun, "Technique and Meaning: The Example of Andean Textiles," *Artforum* 16 (1977): 38–43.

some time after the burial, relatives of the deceased made offerings of cloth and food. These public and ceremonial uses of cloth—often in association with food—were not confined to oneself and one's kin. At rituals of the state religion, cloth and llamas—the latter being a source of coarse wool fibers as well as food—were burned. The Sapa Inca's headdress consisted principally of a cloth cord that he wound around his head and from which there hung cloth fringes that fell over his forehead. Within the Inca administration, a gift of cloth was used to flatter a superior or honor a subordinate. Moreover, the life history of a person articulated with the state through the medium of cloth. Women worked at making cloth for the state as one of the two principal forms of tribute. (Labor in the fields was the other major form of tribute.) The importance of food is self-evident; the fact that food and cloth appear in tandem at crucial phases of Inca life underscores the importance of the latter.

7 Thus far we have discussed spatial arrangement, portability, and the role of cloth. For purposes of exposition, each was treated separately. But the things that characterize Inca insistence form a reticulum; that is, they crisscross in a weblike pattern. Using cloth as a path back into the web, we now pause to consider a few intersections.

The raw materials for making cloth were cotton and wool. Both are easy to transport. Cotton, grown in the lowlands, was freely traded for wool from animals bred and pastured in the mountains. The prepared fibers of either material were spun into yarn with a spindle. The spindle was a stick about ten inches long with a movable circular weight. Women carried them, making yarn as they walked. The yarn was woven into cloth on a simple loom. It consisted of several cords—the warp—stretched between two parallel bars. To prepare the loom for operation, one of the bars was attached to some convenient stationary object; the other bar had a strap that was looped around the weaver's lower back. In operating the loom, the weaver changed the tension on the warp by shifting his or her body. Fiber, spindle, and loom weighed a few pounds and were easily carried in one hand. The simplicity of the weaving tools contrast sharply with the craft of weaving. Compared to all other crafts, it demands the utmost in attention to spatial arrangement. The weaver constructs cloth with thousands of small movements each of which adds yarn (the weft) to the yarn (the warp) stretched between the bars of the loom. Each movement adds a small portion to the design. The overall layout must be kept in mind because small deviations in its execution can dislocate the visual effect of the entire work. Although weavers knew ways to cover over mistakes, full correction was possible only by undoing the cloth back to the place where the problem started.

8 On the desert plains of Nazca, southwest of Cuzco, immense drawings of animals, and long straight lines, are cut into the ground. Set apart from these drawings, but still in the same vicinity, there is a large drawing of a radiating sun. The Incas conquered Nazca; the sun was a principal Inca deity; consequently, the sun drawing at Nazca has been interpreted as a sign of Inca conquest. A more convincing sign of conquest is the sun temple close to the pre-Inca city of Pachacamac. This grandiose structure encloses five acres. Like the sun drawing at Nazca, the temple is set apart. It was built on the top of a conspicuous hill just outside the old walls of the city. But one corner of the Inca temple is only three hundred feet away from the temple built earlier to honor the local Pachacamac deity. Both the Nazca and Pachacamac examples draw on things associated with the state religion. If we were to choose other examples, the same pattern would emerge. The physical signs of Inca conquest were, first of all, highly visible; secondly, native objects were not destroyed; and third, Inca instrusions were placed close to objects that served a similar function in the conquered native cultures. Variations occur in the third part of the pattern. For example, settlements were generally dispersed in the central Andean highlands, so the Incas situated their buildings in locations that would become focal points.

The foregoing pattern is indicative of how the Incas practiced statecraft in newly conquered territories. The visibility of the structures themselves are testimony to their power and authority. On the other hand, local practices continued. For example, the Incas unquestionably controlled agricultural lands. But part of the land was returned to the community. Communal labor was required on adjacent land that was retained by the state and the state religion. Thus a fiction was established in which the Incas returned the land of the community in exchange for work done on state land. In addition to work in the fields and pastures, services were required. This can be seen on the individual level by taking, for example, the work of the potter. With the coming of the Incas, the potter continued to make his wares in the local style, but under Inca influence, he began to produce pottery in the style of Cuzco. For some time during a year, at the behest of the Incas, he might leave the community and pursue his craft elsewhere, but he returned home after his service. In this general scheme, the ayllu community remained intact; local deities continued to be served; and local leaders, under watchful Inca supervision, remained in power. The part of the goods that had formerly been given to members of the community who did not produce them, for example, local leaders and the very old, continued to be given, but now the Incas were included among the beneficiaries.

8 The Nazca drawings are described by Maria Reiche, *Peruanische Erdzeichen* (München: Kunstraum, 1975). The environs of Pachacamac are detailed in W. D. Strong, G. R. Willey, and J. M. Corbett, *Archaeological Studies in Peru, 1941–1942* (New York: Columbia University Press, 1943).

The problem of the influence of the Cuzco style on local pottery is discussed in Donald E. Thompson, "Prehistory of the Uchucmarca Valley in the North Highlands of Peru," *Actas del XLI Congreso Internacional de Americanistas*, vol. 2 (Mexico, D. F.: Comisión de Publicación de Los Actas y Memorias, 1976), pp. 99–106.

Tax service in the form of labor and land division are treated by many authors including S. M. Moore, *Power and Property in Inca Peru*, (New York: Columbia University Press, 1958). Inca patterns of conquest are analyzed by Joseph Bram, *An Analysis of Inca Militarism*, Monographs of the American Ethnological Society, no. 4 (New York: J. J. Augustin, 1941). For a case study of Inca conquest, see Dorothy Menzel, "The Inca Occupation of the South Coast of Peru," *Southwestern Journal of Anthropology* 15 (1959): 125–142. For work on Inca storage, we are indebted to Craig Morris; see for example, Craig Morris and Donald E. Thompson, "Huánuco Viejo: An Inca Administrative Center," *American Antiquity* 35 (1970):344–62; and John V. Murra and Craig Morris, "Dynastic Oral Tradition: Administrative Records and Archaeology in the Andes," *World Archaeology* 7 (1976):267–79.

The formula for Inca conquest was repeated at least forty times. Accordingly, the term *methodical* can very well be used to characterize Inca insistence. Its epitome is found in the Inca storehouse area. These were systematically placed at regular intervals along major roads. Their placement was important. By locating the storehouses as they did, the Incas dropped a meshwork over diverse territories that were the native homes of numerous distinct cultures. The layout of a large storehouse area consisted of hundreds of similarly constructed storage units arranged in rows, together with several service and administrative buildings. The formality of the arrangement is, in itself, indicative of methodological planning and construction. An example of a storehouse area is on the outskirts of Huánaco, a large Inca town about midway on the main road between Cuzco and Quito (plate 3.3). There the storehouses consisted of 497 units arranged in eleven rows (plate 3.4). Most of the units held root crops grown in the surrounding cold, windy pampa.

Plate 3.3. Storehouse area at Huánaco. The storehouses are on the hill. *(Courtesy of the Institute of Andean Research.)*

CODE OF THE QUIPU

The total storage capacity exceeded one million bushels. Thirty additional buildings completed the storehouse area. Some were probably used by people who processed the food stuffs for storage. Other structures were most likely used by people who organized the storage operation, and by those charged with the methodological work of keeping accounts of the produce that entered and left the storehouses.

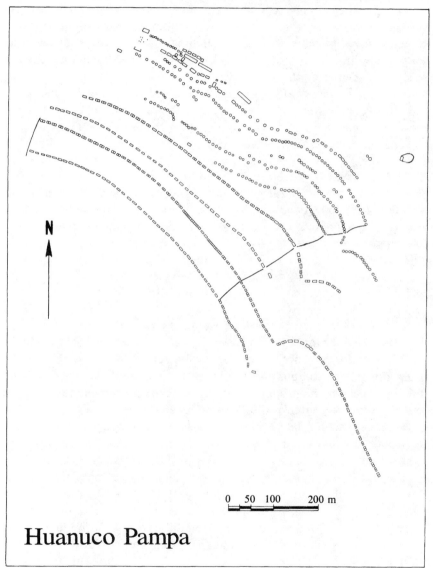

N

0   50   100   200 m

Huanuco Pampa

Plate 3.4. Plan of the storehouses at Huánaco. *(Courtesy of the American Museum of Natural History, New York.)*

9 There are several good sources on Inca architecture. Recent and useful works include Graziano Gasparini and Luise Margolies, *Arquitectura Inka* (Caracas: Universidad Central de Venezuela, 1977); Ann Kendall, "Architecture and Planning of the Inca Sites in the Cusichaca Area," *Baessles-Archiv,* n.s. 22 (1974):73–137; and Roberto S. Wakehan, *Puruchuco: Investigacion Arquitectonica* (Lima: Universidad Nacional de Ingenieria, 1976).

10 Inca masonry has been classified by Wendell C. Bennett in "The Archaeology of the Central Andes," *Handbook of South American Indians,* ed. J. H. Steward, vol. 2, Smithsonian Institution, Bureau of American Ethnology, Bulletin no. 143, (Washington, D.C., 1946), pp. 61–147. Suggestions on how walls were built are in M. K. Jessup, "Inca Masonry at Cuzco," *American Anthropologist* 36 (1934): 239–41; and a discussion of the relationship of those walls to the mountains in the Cuzco area appears in Felipe Cossío del Pomar, *The Art of Ancient Peru* (New York: Wittenborn and Co., 1971). Inty Pata is reported on in Paul Fejos, *Archaeological Explorations Vilcabamba Southeastern Peru,* Viking Fund Publications in Anthropology, no. 3 (New York, 1944). Inca terraces are discussed in O. F. Cook, "Staircase Farms of the Ancients," *National Geographic Magazine* 29 (1916):474–76, 493–534.

9 The Inca technical solution to the problem of storage was not new to the Andes. As with most things, the Incas took an old idea from their past, or borrowed one from their neighbors, and adapted it to meet different ends. The adaptations follow a *conservative* mold that is characteristic of Inca insistence. Faced with the problem of storage over a vast region, they simply copied the same basic unit. When more storage was needed, the number of units was multiplied. However, no basic changes were introduced in the technology of storage. They did the same thing in solving architectural problems. For example, except for scale, the temple of the sun at Cuzco followed the same plan as the residential compound. It consisted of a number of rectangular one room units set around a courtyard arranged within a rectangular, walled enclosure. Access to the compound was through a trapezoidal opening in the enclosure walls. If there was more than one entrance to a unit within the compound, they were spaced at regular intervals. And the trapezoid was programatically repeated for each entrance, the few windows, and the interior wall niches of each unit. In an Inca town, the compound was the basic unit. Ideally, a group of compounds were arranged around a plaza. A town was composed of as many replications of the larger arrangement of compounds with plazas as were required. The notable difference between the architecture of the temple of the sun in Cuzco and the ordinary family compound was the visual effect achieved by the skill and effort put into the construction of the walls.

10 The stone for the enclosure wall of the temple of the sun was cut into rectangular blocks and layed in regular courses. Each of the blocks on the exterior face seems to bulge at its center. Actually, this results from the skillful trimming of the block back toward the edges so that there is an exact and proper fit with the blocks that surround it. The eye travels from the center of the block to its edges, calling attention to the precision of the fit (plate 3.5). The temple of the sun is just one of many important buildings constructed this way. In the walls of still other important structures, many-sided stones of different sizes were similarly worked until they fit together. Some very large stones meet others along six or seven edges, all of them beveled back at different angles. Nothing was used for bonding: the fit of the blocks and the weight of the stone ensured that the wall stood. The Incas did know about bonding materials. The walls of ordinary buildings were made of sun-dried bricks laid in mud, or with unaltered stone set in clay. So the emphasis on exact fit on the exterior faces of what were essentially public buildings was a matter of conscious choice. The walls are forceful visual displays of the Inca's general absorption with fitting everything in its proper place.

Plate 3.5. A girl at an Inca wall in Cuzco, 1975.

Fit characterizes Inca insistence in ways other than architectural finish. It will come up again when we locate the quipumaker within the Inca bureaucracy. For the moment, we give one other appearance of fit, that of man and landscape. The small agricultural settlement of Inty Pata was built on the side of a mountain (plate 3.6). Two stairways fan out and down the mountain (plate 3.7). The principal residential sector was toward the mountain's top. Its walls are gracefully curved in keeping with the mountain's sweep. Below the residential sector, and on both sides of the stairways, there are more than fifty terraces formed by long, curving stone walls that match the natural contours of the site. Behind the walls, a layer of coarse sand and clay was first laid, and then topped with a layer of fine surface soil. Water was conducted down the mountain to the terraces. Access to the crops was via a system of stairs that projected from the terrace walls. From the lower terraces, up to the residential sector, the mountain and settlement rise at a slope that exceeds 40°. The overall impression of Inty Pata is the achievement of a close fit between man's needs and the natural rhythms of the landscape.

Plate 3.6. Inty Pata. The site is at left center. *(Reprinted from* Archaeological Explorations in the Cordillera Vilcambamba Southeastern Peru, *by Paul Fejos.)*

CODE OF THE QUIPU

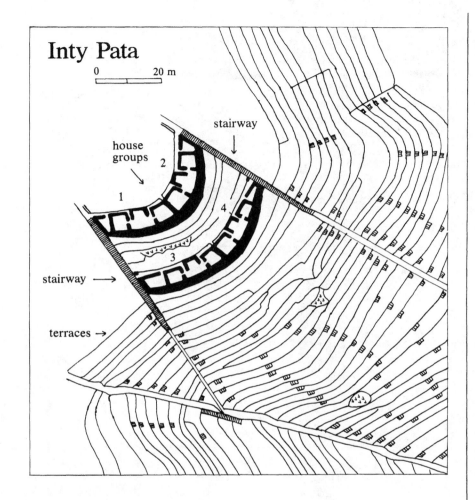

Inty Pata

0   20 m

house groups

2

1

4

3

stairway

stairway →

terraces →

Plate 3.7. Part of the plan of Inty Pata. *(Adapted from* Archaeological Explorations in the Cordillera Vilcambamba Southeastern Peru, *by Paul Fejos.)*

11 Let us focus in on the plan of the residential sector at Inty Pata. It consisted of two groups of two compounds, each of them following the same plan. Each compound has a corner storeroom, two units with small entrances that face each other across the courtyard, and an open unit facing the courtyard. The overall arrangement shows a formality that is marked by repetition. Turning to another area of Inca culture, we note the same kind of thing. Inca pottery decorations repeat the same design over and over again to fill out a band or to cover an area. Rectangles and serrations play an important part in the designs. Design units are structured as rectangles, or rectangles on top of rectangles, and by serrations or diamondlike shapes formed by serrations on top of serrations. These same shapes also appear as elements within the design.

11 Symmetry of Inca Pottery Decorations

(*i*) Examination of Inca pottery leads to the impression of repetition and planning. The patterns are formal and symmetric. In fact, because of the formality, mathematical analysis of the symmetry is possible. The analysis shows that of the variety of types of symmetry possible, the Inca designs are mostly of a relatively few types.

(*ii*) When analyzing a decoration that forms a band or strip, a design unit must first be selected. Repetition is the essence of formal symmetry; the design unit must be repeated as a whole contiguously across the strip. Within the design unit, a smallest basic component is sought. The component is basic in that when it is reflected or rotated it forms the design unit. If there is no reflection or rotation, the basic component is the design unit itself. Most people do recognize as symmetric a design that has reflection across a central vertical line and some refer to this as bilateral symmetry. But, there can also be reflection across a central horizontal line, bifold rotation (through 180°), and reflection across a horizontal line simultaneous with a slide forward. These motions can be applied to the basic component individually or in combination with each other. But, many different combinations of motions lead to the same result and so, no matter how they are combined, only seven different types of design units are possible. Using ⊓ as an example of a basic component, the seven possible design unit types are shown. Associated with each is a brief name to suggest some idea of its formation.

Fig. N.5

⊓	simple repetition	⊓d	bifold rotation
⊓q	vertical reflection	⊓db q	rotation and vertical reflection
⊓ (horizontal reflection)	horizontal reflection	⊓q (double)	double reflection
⊓	horizontal reflection and slide		

There is, or course, no implication that the Inca craftsman conceived of symmetry in this way or that he knew there were seven types and consciously chose from them. This mode of analysis provides a way for us to focus on the structure of the designs and to elaborate on what different looking decorations structurally have in common with each other.

(*iii*) When this mode of analysis was applied to 300 symmetric band decorations, we found that only 13 percent were simple repetition. The six more complicated symmetry types were all present. However, of these, three types were abundant and three were rare. The type of symmetry referred to as *double reflection* alone accounted for 40 percent of the decorations. About another 20 percent were of the *vertical reflection* type and 11 percent were of the type *rotation and reflection*. The other three types together constituted the remaining 16 percent.

Concentrating on the three most abundant types, we can get some insight into how a craftsman might have planned his construction. First, look at the diagrams of a selection of Inca band decorations (fig. N.6). Design units have been indicated by dotted lines and the design type noted.

Fig. N.6

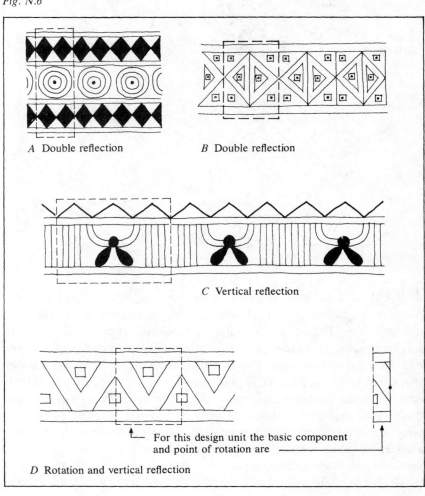

A Double reflection

B Double reflection

C Vertical reflection

For this design unit the basic component and point of rotation are

D Rotation and vertical reflection

If a craftsman intended to repeat a design across a strip, whether or not the boundary lines were to be included, he might conceive of the units this way.

Fig. N.7

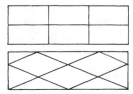

However, design units conforming to this construction plan could not lead to any of the three abundant types. Two construction plans,

Fig. N.8

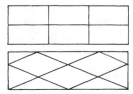 and

seem to have been used by the Inca. Together, they can lead to all of the three types. In many cases, the Inca potter explicitly included the boundary lines.

(*iv*) When decorations cover an area, rather than just forming a band, the movement forward for a band has to be combined with a movement upward. There are also other motions not possible for a band that are possible when a basic component is used to generate a design that repeats to cover a plane. Therefore, different types of symmetry result. For the plane, as in the case of the band, various combinations of the fundamental motions can lead to the same type of design unit. It can be shown that in the plane, no matter how the motions are combined, seventeen distinct types of design units are possible.

(*v*) When 120 symmetric surface decorations were analyzed, the result was striking. Five of the seventeen types of design units were never present and another eight types were present but rare. Four design types predominated and accounted for 70 percent of the decorations. These four predominant design types used only two underlying structures—rectangles on top of rectangles or rhombic cells.

Fig. N.9

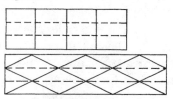

Notice that the structures underlying the abundant band design types can simply be placed one on top of the other to achieve these.

Fig. N.10

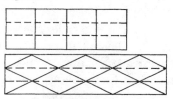

The design within the cells also show that the Incas did not expand their decoration vocabulary when constructing a surface decoration rather than a band decoration. Two of the four predominant design unit types have both vertical and horizontal reflection and a center of bifold rotation. (Examples are shown in fig. N.11.) The other two types have a single line of reflection. As with the bands, double reflection is the most abundant and single reflection is next.

Fig. N.11

Thus, the symmetry of Inca pottery decorations is characterized by repetition of designs based on the rectangle, serrations, and double or single reflection.

(*vi*) The analysis of symmetry on bands is found in Anna O. Shepard, *The Symmetry of Abstract Design with Special Reference to Ceramic Decoration,* Publication 574, Contributions to American Anthropology and History, vol. 9, no. 47 (Washington, D.C.: Carnegie Institution of Washington, 1948).

The decorations analyzed are those in Jenaro Fernández Baca, *Motivos de Ornamentación de la Cerámica Inca-Cuzco,* vol. 1. (Liberia Studium, S.A., 1971). The decorations in figure N.6 are (*A*) no. 46, (*B*) no. 490, (*C*) no. 126, and (*D*) no. 249 in Baca's volume. Those in figure N.11 are (*A*) no. 615, (*B*) no. 103, and (*C*) no. 561.

Plane symmetry is discussed in Doris Schattschneider, "The Plane Symmetry Groups: Their Recognition and Notation," *American Mathematical Monthly* 85 (1978): 439–50.

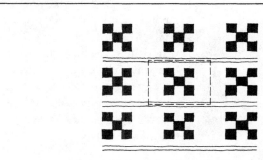

A Rectangular cell with double reflection

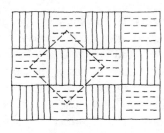

B Rhombic cell with double reflection

C Rhombic cell with double reflection

Another important feature of the designs is the use of mirror images. Within the design units, there are reflections over a central vertical line, or both reflections over a central vertical line and central horizontal line. (Examples are shown and discussed in the chapter notes.) At Inty Pata, note that one row of compounds is repeated above the other, and within each row there is a reflection over a central line. Some or all of these features, evidenced in pottery decorations, compound arrangements, storehouses, fabric designs, the system of ayllus, the walls of the temple of the sun in Cuzco, and elsewhere, define the particular kind of *symmetry* in a strict sense that characterizes Inca insistence.

12 On flutes see C. W. Mead, *The Musical Instruments of the Incas*, American Museum of Natural History, Anthropological Papers, vol. 15 (New York, 1924), pp. 313–49.

12 A coda needs to be added to the description—now completed—of things that characterize Inca insistence. Our point of departure has been artifacts. Inca artifacts also show the variety that results when individuals, with plans in mind, working with their hands and a few tools, produce particular concrete objects. Take the flute for example. The plan was simple: a hollow tube with some holes and a notch at one end. Actual flutes vary fivefold in length, and twofold in the number of holes. The spacing of holes along the body is different on each instrument. Some Inca flute makers apparently experimented with hole spacing. On several instruments, one or more of the original holes were permanently plugged, and new holes were drilled at different places. Given the variety in the length of the flute body, the number of holes, and their spacing, no two instruments sounded exactly the same. Yet, in spite of the differences, the haunting music of the flute was produced by every instrument. What is true for flutes holds true for the realization of plans of any kind. It is particular people in particular situations who carry out plans, whether it be governing a conquered territory or making a musical instrument.

Cotton and wool cloth, portability, methodical, concern for spatial arrangement, symmetry, conservative, and fit, are the things that characterize Inca insistence. The quipu is their quintessence. Quipus were made from cotton or wool. If we did not know the important role of cotton and wool fabrics in Inca culture, this would seem a highly unlikely material for a medium. On the other hand, few media before modern times were as portable. Most quipus weigh less than a pound. They can be bunched up, folded up, rolled up, or otherwise compacted, without the slightest damage. Both the construction and contents of quipus reflect the systematic and orderly approach which is the mark of methodical people. A quipu is a composition of multicolored cords in space. The placement of the cords relative to each other—that is, their spatial arrangement—carries a large portion of the stored information. As with Inca symmetric forms, this spatial arrangement uses formal repetition and recombination of basic elements with emphasis on rectilinearity. The quipus we know about all date from Inca times. However, all are equally developed so the general idea almost certainly derived from tradition founded in earlier phases of Andean history. They fit the ethos of the Andes, were adapted to the particular needs of a conquest state, and are its metaphor.

Chapter **4** | # The Quipumaker

1 What particular abilities did a person need to be a quipumaker? What was his position in the Inca bureaucracy? In what ways did one quipumaker differ from another? And what did the quipumaker have to know? These are interesting questions, and they are going to be answered. The route to the answers will often appear to be a thin line of scant information, or a dotted line—information with many gaps—or a broken line as when, for example, information from another culture is introduced. At the end, all questions having been answered, there will emerge a still too darkly shaded picture of the quipumaker.

2 His material—colored strings of cotton and sometimes wool—will give us some notion of the abilities the quipumaker needed. They become apparent when we contrast his material with those of his counterparts in other civilizations.

Many different substances have been used for recording. Stone, animal skin, clay, silk, and various parts of plants including slips of wood, bark, leaves, and pulp are some of them. The material used for a medium in a civilization is often derived from a substance that is common and abundant in its environment. (The simultaneous use of several mediums in one area is a recent development. Even if two were used, one tended to dominate the other and gradually replaced it.) Each kind of material calls forth a somewhat different set of abilities. For contrast with the quipumaker's cotton and wool, we choose to detail the clay of the Sumerian scribe and the papyrus of the Egyptian record keeper.

The Sumerian scribe lived in the southern part of what today is called Iraq, between, say 2700 and 1700 B.C.E. The clay he used came from the banks of rivers. He kneaded it into a tablet that varied in size from a postage stamp to a pillow. (For special purposes, the clay was shaped into a tag, a prism, or a barrel.) Pulling a piece of thread across the clay, he made rulings on the tablet. He was then prepared to record. This he did with a stylus, a piece of reed about the size of a small pencil shaped at one end so that it made wedgelike impressions in the soft, damp clay. If he lived toward the early part of the thousand-year time interval, he made impressions vertically, from top to bottom. Later on, they were made from left to right across the tablet. Having finished one side, he turned the tablet over by bringing the lower edge to the top, continuing the record on the obverse side. He had to work fast: the clay dried out and hardened quickly; when that happened, erasures, additions, and other changes were no longer possible. If he ran short of space on the tablet, or if the tablet dried out before he was done, he started a second one. When he was finished recording, the tablet or tablets were dried

2 A general discussion of recording materials appears in David Diringer, *The Hand-Produced Book* (London: Huchinson's, 1953). The importance of media in civilization is carried further than anywhere else in Harold Innes, *Empire and Communication* (Oxford: Oxford University Press, 1950). His notion that media light in weight go with civilizations that stress space over time is supported by the Inca case.

An introduction to Sumerian civilization is Samuel N. Kramer, *The Sumerians* (Chicago: University of Chicago Press, 1963); and Edward Chiera, *They Wrote on Clay* (Chicago: University of Chicago Press, 1938), is a good source on the making of clay tablets. There are many general books on Egyptian civilization; a dependable text, written in 1923, and translated from the German, is Adolf Erman, *The Ancient Egyptians,* (New York: Harper and Row, 1966). The definitive work on papyrus is Naphtali Lewis, *Papyrus in Classical Antiquity* (Oxford: Oxford University Press, 1974). What the Spanish chroniclers had to say about education during the time of the Incas, including the education of possible quipumakers, is summarized in Daniel Valcárcel, *Historia de la Educación Incaica,* (Lima: Banco Comercial del Perú, 1961). For comparative, specific information on the Sumerian scribe, see Samuel N. Kramer, "School days: A Sumerian Composition Relating to the Education of a Scribe," *Journal of the American Oriental Institute* 69 (1949): 199–215; and for Egypt, see W. K. Simpson, ed., *The Literature of Ancient Egypt,* (New Haven: Yale University Press, 1973), particularly the sections called "The Instruction of

Amunnakhte'' and ''The Instruction of Dua-Khety.''

Interesting observations on the feel of things, for example wood as compared with plastic, are written about in Roland Barthes, *Mythologies* (New York: Hill and Wang, 1972). See, in particular, the essays ''Toys'' and ''Plastic.'' A cogent statement on the sense of touch is found in Lawrence K. Frank, ''Tactile Communication,'' *Explorations in Communication,* ed. E. Carpenter and M. McLuhan (Boston: Beacon Press, 1960). The art historian who discusses Andean textiles in terms of music is George Kubler; see his *Art and Architecture of Ancient America* (London: Penguin Books, 1962).

in the sun or baked in a kiln, permanently fixing the impressions. A tablet by a Sumerian scribe is shown in plate 4.1; later, we return to its contents.

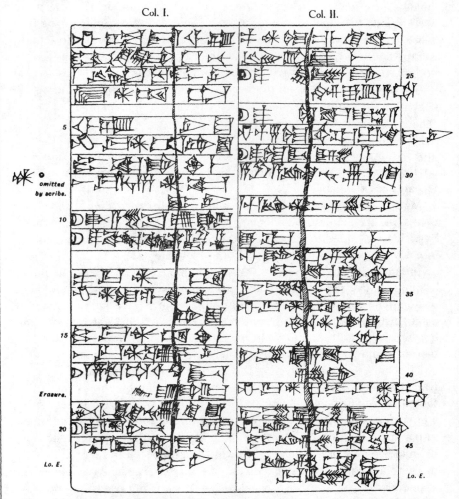

Plate 4.1. Sumerian clay tablet. *(Reprinted from* Ur Dynasty Tablets, *by J. B. Nies. Courtesy of the publisher.)*

In Egypt, at about the same time, the scribe used papyrus. Its source was the interior of the stem of a tall sedge that flourished in swampy depressions. Fresh stems were cut, the rinds were removed, and the soft interiors were laid out and beaten until they were formed into sheets. The natural gum of the pith was the adhesive. A papyrus sheet was about six inches wide and nine inches high. It was white or faintly colored, the surface was shiny and smooth, and it was flexible. Dry

CODE OF THE QUIPU

sheets could be joined with a prepared adhesive; twenty of them, for example, made a surface six feet long. The Egyptian scribe used brush and ink. To make a brush, he cut a rush about one foot in length; then, he cut one end at an angle and bruised it to separate the fibres. His inks were actually small cakes resembling modern water colors and they were used in much the same way. Black cakes were made with soot scraped from cooking vessels; red cakes, from ocher. Moving from right to left, the Egyptian scribe brushed his record onto the papryus.

An obvious contrast between the quipumaker and his Sumerian and Egyptian counterparts is that the former used no instruments to record. The quipumaker composed his recording by tracing figures in space as when, for example, he turned a string in an ever changing direction in the process of tying a knot. All of this was not preparatory to making a record; it was part of the very process of recording. The stylus and the brush were held in the hand, their use had to be learned, and the learning involved a sense of touch. But the quipumaker's way of recording—direct construction—required tactile sensitivity to a much greater degree. In fact, the overall aesthetic of the quipu is related to the tactile: the manner of recording and the recording itself are decidedly rhythmic; the first in the activity, the second in the effect. We seldom realize the potential of our sense of touch, and we are usually unaware of its association with rhythm. Yet anyone familiar with the activity of caressing will immediately see the connection between touch and rhythm. In fact, tactile sensitivity begins in the rhythmic pulsating environment of the unborn child far in advance of the development of other senses.

Color is another point of contrast: the Sumerian used none, the Egyptians two (black and red), and the quipumakers used hundreds. All three needed keen vision; the quipumaker alone had to recognize and recall color differences and use them to his advantage. His color vocabulary was large; it was not simply red, green, white, and so on, but various reds, greens, and whites. Drawing upon this color vocabulary, his task was to choose, combine, and arrange colors in varied patterns to express the relationships in whatever it was that he was recording. Confronted with a quipu, it is not easy to grasp immediately, if at all, the complex use of colors. The quipumaker, and the people of the Andean world who were a part of his everyday experience, understood complex color usage because they were accustomed to it in the textiles they saw, just as we comprehend polyphonic music because we hear it often enough. This appeal to musical imagery comes from an art historian; others in our culture have also turned to musical composition to translate their understanding of Andean color composition. At the base of their musical imagery is the formal patterning and structure which can also be translated into mathematical language.

The third contrast is perhaps the most important. Both the Sumerian and the Egyptian recorded on planar surfaces. In this regard, papyrus had certain advantages over clay. For example, sheets could be added or deleted, thus changing the dimensions of the surface; the dimensions of the clay surface were fixed once the tablet was formed. By contrast to both papyrus and clay, the quipumaker's strings present no surface at all. Recording in papyrus or clay involved filling the space in a more or less continuous process either up or down, or from right to left, or from left to right. This is linear composition. By contrast, the quipumaker's recording was nonlinear. The nonlinearity is a consequence of the soft material he used. A group of strings occupy a space that has no definite orientation; as the quipumaker connected strings to each other, the space became defined by the points where the strings were attached. The establishment of these points did not have to follow any set left-to-right or right-to-left sequence. The relative positions of the strings are set by their points of attachment, and it is the relative position, along with the colors and the knots, that render the recording meaningful. Essentially then, the quipumaker had to have the ability to conceive and execute a recording in three dimensions with color.

3 The quipumaker fits somewhere in the bureaucracy that developed in the Inca state: the question is, Where? In theory, his position was one of privilege. As for the facts in the case, the one good piece of evidence that exists supports what one would expect from theory.

Hand in hand with massive construction, standing armies, and all the other attributes of the state, there is always a bureaucracy to administer its affairs. And bureaucratic administration, in the words of Max Weber ". . . means fundamentally the exercise of control on the basis of knowledge." The knowledge is stored in records. These, together with people who have "official functions," form the "office" that carries on the state's affairs. The bureaucracy keeps records of everything that can be recorded, but especially things that are quantifiable: the number of people living at a certain place, the tribute that was collected in a village, the day the river flooded. The bureaucracy believes in its rationality; its records give assurance to those who wield power. The more records there are, and the more the bureaucracy has experience with them, the more power to the state. A bureaucracy's records are peculiar to itself, and bureaucrats try very hard to keep it that way.

In the Inca state, the quipumaker composed the records for the bureaucracy. He might know, for example, how many men in a group of

3 Max Weber was, and remains today, the person who best understood the nature of bureaucracy. His generalizations are based on studies that range from early civilizations to modern corporations and they apply with equal force in all cases. The thoughts that we use come from his "The Essentials of Bureaucratic Organization: An Ideal-Type Construction," *Reader in Bureaucracy,* ed. R. K. Merton et. al (Glencoe, Ill.: The Free Press, 1952); and, *From Max Weber: Essays in Sociology,* ed. H. H. Gerth and C. Wright Mills (Oxford: Oxford University Press, 1946). For a general discussion of the scribe in bureaucracies of early civilization, nothing has come along to surpass V. Gordon Childe, *Man Makes Himself* (1934; reprint ed., New York: Mentor Books, 1951). (Incidentally, Herman Melville's story, "Bartleby the Scrivener," might well be read in connection with this section of the chapter. It is available in several editions; originally published in 1853, there is now a facsimile reproduction in Howard P. Vincent, *Bartleby the Scrivener* (Kent, Ohio: Kent State University Press, 1966).

villages were suitable for army service, how many could work in the mines, and much else of interest. He worked with privileged information, so he was privileged. We expect that he was more important than an ordinary man, yet he was not as important as the really important men who held authority in the community where he lived, or the Incas who watched over them. The evidence from a cemetery bears out the expectation.

All quipus that we know about were taken from cemeteries where they had been buried with dead persons for whom they had meaning in life. Graves with quipus in them also contained other objects; and close by, there were graves where people were buried with objects important in their own lives. Unfortunately, those who took quipus from cemeteries destroyed the contextual relationship. Things of value were kept, the remainder were destroyed. Valuable objects were thoroughly mixed without regard to which burial, or even what cemetery they came from. No notes were taken. As a result, almost all the quipus now in museums stand completely apart from their association with other related objects and persons. There is an exceptional case. In 1900 and 1901 Max Uhle took quipus from a cemetery; he also took notes, and he preserved many of the objects. These notes and objects, carefully reevaluated by Dorothy Menzel, and examined by us, are the evidence. The cemetery is on the outskirts of Ica, a town in a valley with the same name.

Ica is in the coastal desert southwest from Cuzco. The Incas moved upon this area at the start of the last quarter of the fifteenth century. When they arrived, Ica was already a thriving place. As elsewhere, the Incas did not destroy what they found: under the conquerors, Ica became an administrative center and a key way station on the main coastal road. The old Ica cemetery continued to be used, but now the contents of the burials showed Inca influence. For example, the pottery in graves retained a local touch, but much of it was inspired by Inca styles, or direct imitations of them. Two burials from the Inca period contained quipus; Uhle's notes on one of them are clear enough to permit interpretation. The contents of this burial will be compared with the contents, also based on Uhle's notes, of the burial of an ordinary man and a very important person.

The ordinary man was buried in a sitting position inside a burial urn. He wore a shirt and a blue cap. Shell beads, a weaving sword, and pottery were in the urn with him. The very important person was placed in a small roofed structure, something like a room built into the ground. Two other dead persons, similarly wrapped, were seated with him. There were burial urns in front of the dead; in back of them stood pottery vessels; and in back of that, there were elaborately carved wood imple-

Max Uhle's notes are deposited at the Lowie Museum of Anthropology, University of California, Berkeley. In his notes, the grave of the person we call "the ordinary man" is designated Th-2; our "important man" is Td-8; and the burial with the quipus is Tk. The careful scholarship in Dorothy Menzel, *Pottery Style and Society in Ancient Peru* (Berkeley and Los Angeles: University of California Press, 1976), is noteworthy; she includes insightful observations on the materials that Uhle removed from graves in the vicinity of Ica; her main concern is the pottery and its interpretations within a cultural context.

The letter that Guamán Poma wrote to the king appears in a facsimile edition: *Nueva Crónica y Buen Gobierno* (Paris: Institut d'Ethnologie, Université de Paris, 1936). The facsimile is difficult to read; there is a conversion of Poma's handwriting and grammatical usages into print and modern Spanish: see Coronel Luis Bustíos Gálvez, *La Nueva Crónica y Buen Gobierno* (Lima: Gráfica Industrial, 1966). The best general description of Poma's work is John V. Murra, "Guamán Poma de Ayala," Pt I, II, *Natural History* 70 (1961). A literary analysis of Poma is: Rolena K. Adorno, *The Nueva Crónica y Buen Gobierno: A Lost Chapter in the History of Latin-American Letters* (Ann Arbor: University Microfilms International, 1974). There are three other drawings in addition to those reproduced here where Poma shows people holding quipus. In the facsimile edition, the drawings are on pages 201, 800, and 883. Poma also mentions quipus in passing elsewhere in the text of his letter. However, none of these are related to how Poma depicted quipus within the Inca bureaucracy. The object in the lower left of Plate 4.3D is a counting board. It is unclear from Poma's commentary whether it is his version of a device associated with Spanish activities analogous to those of the person depicted or whether he is implying its association with the Incas. In either case, his commentary makes interpretation of the configuration and the meaning of the unfilled and filled holes highly speculative. For further discussions of it see H. Wassen, "The Ancient Peruvian Abacus," *Comparative Ethnological*

Studies 9 (1931): 191–205; L. L. Locke, "The Ancient Peruvian Abacus," *Scripta Mathematica* 1 (1932): 37–43; C. L. Day, *Quipus and Witches Knots* (Lawrence: University of Kansas Press, 1967); and our critique of these in Marcia Ascher and Robert Ascher, "Numbers and Relations from Ancient Andean Quipus," *Archive for History of Exact Sciences* 8 (1972): 288–320.

Franz Fanon wrote several books on colonialism. All stress its dire effects on the person. His most general work is *The Wretched of the Earth* (New York: Grove Press, 1968).

ments sheathed with gold and silver. There was also a stool. The wood implements were traditional in the burials of very important persons in Ica before and after the Inca. The stool signified a person with local authority under Inca domination.

No special room was built to receive the bodies and objects of the burial that included the quipus. In this regard, it was like the grave of the ordinary man. The quipu burial contained oversized carved wood implements, but they were not sheathed in gold and silver. In some respects, the burial with the quipus was like that of an ordinary man; in others, like that of a person of moderate importance in Ica. What is most telling are the objects that appear in the quipu burial and nowhere else in the entire cemetery containing more than thirty burials from the Inca period. They are Inca pottery made in Cuzco, distinctive Inca wood drinking cups, and pottery imported from an area north of Ica.

The evidence can be interpreted this way. With an expanding state, more bureaucratic functionaries were in demand. To meet this demand, the Incas drew upon people in conquered areas who had some standing at home. The person in the burial with quipus was a native of Ica who lived for a while in Cuzco where he learned Inca ways in general, and the craft of quipumaker in particular. Then he returned home to serve as a bureaucrat in the Inca administration. He was a privileged person. Other interpretations of the evidence are possible. Although they might differ in detail, all would be in accord with what we would expect from theory.

Perhaps one century after the burials just described, and about eighty years after the Spanish conquest of the Incas, Felipe Guamán Poma de Ayala addressed a 1,179 page letter to the king of Spain. We introduce some of his views by way of a commentary on our question.

Poma was a native Andean who grew up under Spanish colonialism. As a young man he traveled on church business. This much is known about his life because he tells us so in his letter. More is known about his views. Poma wrote to the king to inform him of abuses under the Spanish. But first he puts things in perspective. He begins with the Holy Trinity and the Creation of Adam and Eve; then, he introduces his family and himself. This is followed by the Five Ages of the World, the establishment of the Church, and the discovery of America. The remaining introductory section, about one-third of the entire manuscript, presents his view of Andean life, and particularly of the rule of the Incas. Then Poma gets to the point: the rest of the letter is about corruption, hypocrisy, cruelty, and torture introduced by the Spanish. He concludes with a plea for better government and true Christianity.

As a statement on the abuses of colonialism, the letter has few equals.

CODE OF THE QUIPU

But Poma, after all, suffered from the abuses he writes about. He was taught to put aside his heritage, but he found things in it that gave his life dignity; a nonperson to the Spanish, he nevertheless thought that their religion, and an improvement in their leadership, was an answer to his problems and the problems of the Andes. The generally idyllic section on the Incas has internal contradictions, imaginative chronologies, and other difficulties; the section on the colonial period is thoroughly ambivalent. Nevertheless, Guamán Poma tells us how the Incas appeared to a talented "colonized man" (read Franz Fanon) three generations removed.

In one of the remarkable 397 drawings that are full pages in the letter, Poma imagines himself presenting his work to the king (plate 4.2). In

Plate 4.2. Reprinted from Nueva Crónica y Buen Gobierno, *by Felipe Guáman Poma de Ayala.*

DEPOCITODELINGA
COLL CA

ADMINISTRADORDEPROVINCIAS
SVIVIOC·GVAIAC·POMA

SECRETARIO·DELINGAICONSEJO
INCAP·QVIPOCNINCAPAC
APOCOMARCAMACHICVINIMQVIPOC

CONTADORMAIORITESORERO
TAVANTIN·SVIOQVIPOC
CVRACA·CON DOR·CHAVA

Plate 4.3. Reprinted from Nueva Crónica y Buen Gobierno, *by Felipe Guáman Poma de Ayala.*

one section of his letter, there is a series of illustrations devoted to the Inca bureaucracy. Four of these show men with quipus (plate 4.3). The writing on the drawings, as on others in the text, follow a similar scheme. The top line is Spanish; below is the Quechua equivalent. The people in the drawings are often named. And at the bottom of the drawing, Poma usually has a caption. For example, in 4.3*A*, DEPOCITODE-LINGA and COLLCA, translate into Inca storehouse, and storehouse, respectively; the script on the left names a particular Sapa Inca, topa ynca yupanqui; on the right, in mixed Spanish and Quechua, Poma has written administrador suyoyoc apo poma chaua. This identifies the man displaying the quipu as a very important functionary, and also a claimed ancestor of Guamán Poma. The caption on the bottom repeats the top line in the plural; that is, Inca storehouses.

The man shown with a quipu in each hand (4.3*B*) is also a suyoyoc, who was, says Poma, the administrator of a province. As such, he was responsible for many things, including the records of the communities under his charge. The position was hereditary, but not automatically so; to follow his father, a son had to show diligence and efficiency. In drawings 4.3*C* and 4.3*D*, the scene switches from the conquered territories to Cuzco. The person in 4.3*C* is no less than the secretary to the Inca and his council; and the man in 4.3*D* is the chief treasurer and accountant. In the texts accompanying both of these drawings, Poma points out that secretaries and treasurer-accountants were in each city, town, and village. And to help the king understand, in both cases Poma compares their work to what the Spanish do with pen and ink.

Now in Poma's visual vocabulary, objects held in the hands convey unequivocal messages. Here the objects are quipus. But just about everyone in the 397 drawings is holding something. Priests hold a bible in one hand and a crucifix in the other; soldiers grasp a spear and a shield or a club; and women are depicted spinning or holding flowers. For example, Guamán Poma holds a book in his hand; the king, his scepter. Even empty hands are busy: Poma and the king gesture with their free hand. The quipus in the drawings are not necessarily made by the people holding them—particularly so for the two suyoyocs—any more than the king made his scepter. Poma is visually associating high level functionaries with quipus: just as the soldier has his spear, the priest his crucifix, and the king his scepter, so bureaucrats have their quipus. In a sense, Poma's bureaucracy is the same as Max Weber's.

4 Thus far, the term the quipumaker has been used in the same way one might say the lawyer or the taxi driver meaning people who do the same kind of work. But individual quipumakers surely differed with

4 (*a*) The map in plate 4.4 includes the major sites where Inca quipus have been found. For a list of the current locations of known quipus see Marcia

Ascher and Robert Ascher, *Code of the Quipu Databook* (Ann Arbor: University of Michigan Press, 1978), available on microfiche from University Microfilms International. The map, the databook, and our considerations exclude cords with knots currently used in the Andes. The contemporary cords with knots differ fundamentally in concept and cultural context. For further discussion of them, see: Max Uhle, "A Modern Quipu from Custusuma, Bolivia," *Bulletin of the Free Museum of Science and Art* 1 (University of Pennsylvania, 1897): 51–63; C. J. Mackey, *Knot Records in Ancient and Modern Peru* (Ann Arbor: University Microfilms International, 1970); Enrique de Guimaraes, "Algo Sobre El Quipus," *Revista de Histórica* 2 (1907): 55–62; and Froilán Flores, "Los Kipus Modernos de la Comunidad de Laramarca," *Revista del Museo Nacional* 19 (1950).

(*b*) The quipus that were contained in the cloth bag are AS59–AS67. Both forms of top cords are used on AS66. This set of quipus has also been discussed by C. J. Mackey (*Knot Records*). The quipu sets for which color construction is contrasted are AS147–AS149 and AS33 A–G. The quipus with top digit alignment are AS146 and AS164. Fig. 4.1*A* is AS195 and fig. 4.1*B* is AS194. Figure 4.2*A* is AS159. Figure 4.2*B* is AS173. Figure 4.2*C* is AS174. Plate 4.6 shows AS33. The quipus suspended from carved bars are discussed in further detail in section 10 of chapter 5. One of them is quipu example 5.7. The set of large quipus are AS69–AS71. Two of them are discussed in further detail in section 5 of chapter 6 as quipu example 6.4 and quipu example 6.6.

respect to their work just as taxi drivers and lawyers do. The question is, How did they differ? The only way to answer the question for quipumakers is to look at the products of their work. To do this, we first have to be able to identify sets of quipus made by individuals.

This is easier said than done. There are now about 400 known quipus in the world. This is a strikingly small sample. After all, a single quipumaker, producing say one quipu a week for eight years, would have made more than there are in the entire world sample. But remember that quipus have come from graves. The quipus buried with a quipumaker are those few that just happened to be in his possession when he died. So the first problem is that quipus that can be associated with an

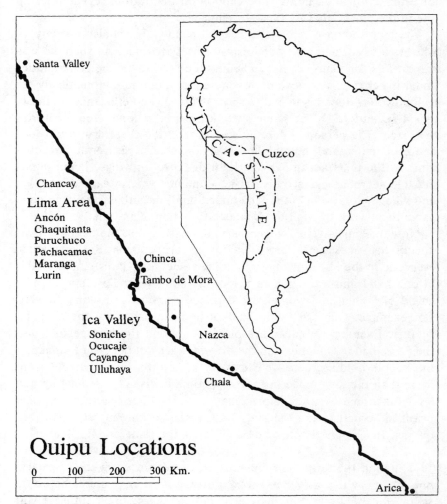

Plate 4.4. Quipu locations.

CODE OF THE QUIPU

individual are a small and biased sample of his workmanship. The second problem is a consequence of the fact that all quipus have come from graves in the desert. (The seventeen locations are shown in plate 4.4.) It is in the desert that conditions are good enough for the preservation of cotton and wool. We do not know how many quipus are still below

Plate 4.5. A quipumaker's bag. *(In the collection of Oscar Núñez del Prado, Cuzco, Peru.)*

the ground, but it is unlikely that they will be discovered in mountainous, wet regions. This tells us that whatever the individual differences, they are limited to those among quipumakers who died far from rainy Cuzco, the center of Inca power. The third problem is the biggest block in the way of answering the question. After removal from graves, quipus were not only separated from persons and artifacts, as stated above, but they were often separated from each other. Some of those that were not discarded were dispatched individually, or in mixed bundles, to museums spread across three continents. Once there, the museums, not knowing exactly where the quipus came from, added them to other quipus in their collections. It would be impossible to associate one set of quipus with one quipumaker if this had happened in every case. There are instances where we know, or strongly suspect, that a set of quipus was made by one person. We draw upon these to answer the question.

Quipumakers differ from each other in the way that no two people write alike. Quipumakers also differ in the way that some people write more legibly than others. For example, no two quipumakers made knots exactly the same, and some took more care than others in the formation and placement of knot clusters. The differences point up individuality in workmanship and give a sense of person, but better expressions of it are found in the exercise of self-conscious choices. One example of this comes from a cloth bag (plate 4.5) which contained two bundles of prepared pendant cords and a set of eight quipus. In explaining how to make a quipu, two equally possible ways of forming top cords were given (see fig. 2.5). The quipumaker whose bag this was, chose to use both ways, and, uniquely, he chose to use them both on one quipu. The inclinations of different people are shown in the way they elect to combine colors. In explaining how to make a quipu, three ways were presented to construct new colors from already dyed yarn (chap. 2, section 6). One quipumaker used one kind of construction to produce four colors and a second kind to produce a fifth color. A fellow quipumaker, also interested in producing five colors, used the first construction for two colors, and the second to make three colors. And, finally, there is the case of the rather imaginative quipumaker who showed volition in the way he placed knot clusters when recording numbers. Normally, knot clusters for the units position are aligned from cord to cord. He chose, instead, to align the positions of highest value. This would be analogous to our lining up the leftmost digits in a column of figures rather than the rightmost digits; it is as if we were to write

$$\left. \begin{array}{l} 1564 \\ 93 \\ 427 \end{array} \right\} \quad \text{rather than} \quad \left\{ \begin{array}{l} 1564 \\ 93 \\ 427 \end{array} \right.$$

A set of quipus made by one person is often conceptually related. It is as if an individual at the time of his death was working on a single problem that necessitated the creation of several quipus. We extrapolate backward and suggest that a quipumaker often worked this way. Figure 4.1 is a schematic of two complete quipus. At first glance, they look identical. The cord layouts are the same and so are the values, accounting for the initial reaction. A closer look brings out the differences: figure 4.1A has an additional subsidiary, and it ends in a bulb, while figure 4.1B ends in a cord placed at a good distance from the pendant. The color signing differs: the main cord in 4.1A is a mix of two colors while the main cord in 4.1B is a mix of four colors; and in three places, where 4.1A has color C1, 4.1B has color C4. Finally, the order of two subsidiaries is reversed. The quipus are not identical, yet they are cer-

Fig. 4.1

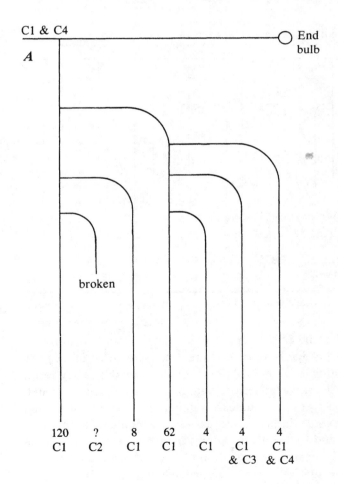

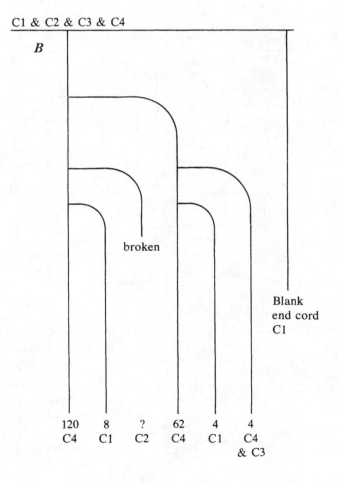

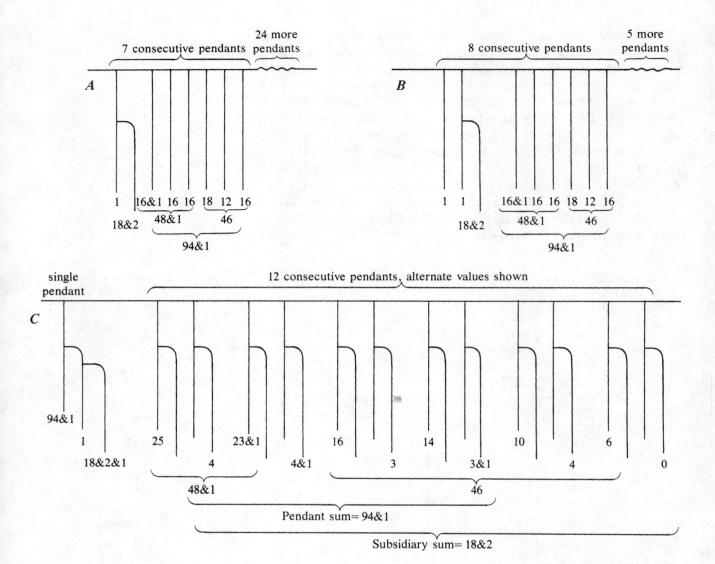

Fig. 4.2

tainly close enough to be called two versions of one record. A different kind of conceptual linkage is seen in figure 4.2. Here three quipus are involved. They distinctly differ from each other in layout. They refer, but only in part, to the same numerical data. Some numbers are repeated from quipu to quipu, but other numbers summarize or recombine those on another quipu. Notice in figure 4.2 that related numbers on all three quipus are consistantly placed on either pendant or subsidiary cords and in either the higher or lower position of a cord with multiple numbers. Clearly, these quipus are related around one idea. In some in-

CODE OF THE QUIPU

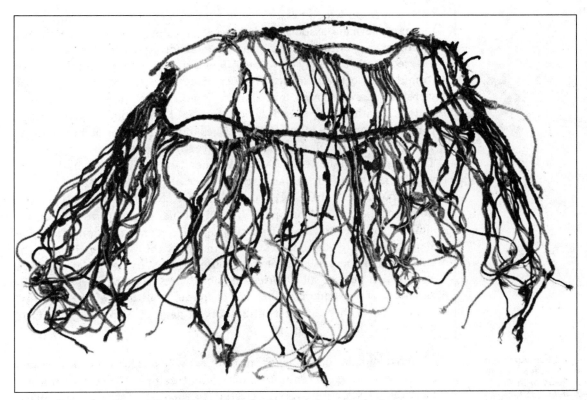

Plate 4.6. Seven quipus tied together. *(In the collection of the Peabody Museum, Cambridge, Massachusetts.)*

stances, a quipumaker tied a series of related quipus to each other. This, no doubt, made it easier for him to keep them together; it certainly makes our task easier. An example is shown in plate 4.6. There are actually seven quipus in the photograph; three were tied to form a circle, and four more were hung from these three. This set of quipus can be summarily described as a rolling series involving colors, construction features, and numbers that take on particular prominence.

Quipumakers who can be identified as persons may have worked at any number of places, say at a building site, or with armies on the move. It seems likely that some quipumakers were in charge of, or responsible for, the work of their fellows. Given the wide range in sets of quipus, it is tempting to say that some people worked in the arena of the ayllu, the smallest traditional organizational unit, whereas others were attached to the administration of larger units within the organization of the Inca state. Put another way, quipumakers were privileged persons, but some were more privileged than others. The temptation to draw conclusions on which persons were which comes from a comparison of the paraphernalia associated with a set of quipus and from the size of the quipus in different sets. The quipumaker who kept his work in a tattered cloth bag may have been less privileged than another whose formidable work was intricately suspended from carved wood bars that were the mark of importance in the region where he died. (Compare plate 4.5 with plates 5.1, 5.2, 5.3, and 5.4.) In a similar vein, let us compare two people in terms of the sizes of the quipus buried with them. The quipus of the first person have already been shown in figure 4.1. There are two in the set, each with one pendant cord. The second person was buried with four quipus. The number of pendant cords on each of his quipus are: 80, 102, 251, and over 2,000. Returning to the first person, note that the pendant on one of his quipus (fig. 4.1B) carries three subsidiaries on the first level, and two on the second. By comparison, a single pendant of the smallest quipu made by the second person might carry as many as six levels of subsidiaries; that is, on one pendant, there are subsidiaries, of subsidiaries, of subsidiaries, of subsidiaries, of subsidiaries, of subsidiaries. The size of quipus and the paraphernalia associated with them suggest differences in degrees of privilege, but we cannot say that for sure.

5 What did the quipumaker have to know? Recall that in his travels Cieza de León gathered information about quipus. Cieza does not answer the question for us, but we return to what he wrote as a good place to start to find an answer.

Cieza devoted one short chapter to quipus. More on the subject is scattered throughout his chronicle. On two occasions, he tells us about censuses kept on quipus. Some categories in the census are given, but he settles for the general notion rather than a full explanation. In like manner, Cieza writes about accounts kept on quipus of goods entering or leaving storehouses; again, full exposition is lacking, as it is in places where he mentions quipu accounts of tribute in the form of goods for the Inca. In three places, Cieza writes about songs, ballads, and lays concerned with the deeds of Sapa Incas. He associates quipus with these state histories in song, but the connection is not clear. Nothing is gained by dwelling on what Cieza could not, or did not, write. It is important that he tells us that quipus served in many different contexts.

Chroniclers who followed Cieza in time add to the list of contexts in which quipus were apparently used. Cristóbal de Molina writes about quipus and calendrics, Garcilaso de Vega adds the context of law, and Bernabé Cobo links quipus with peace negotiations. These three and others, for example, José de Acosta and Polo de Ondegardo, echo Cieza's

5 For analysis of oral literature, and the selections used here, we draw upon Jan Vansina, *Oral Tradition: A Study in Historical Methodology* (Chicago: Aldine, 1965), source for the Burundi narrative; and Albert B. Lord, *The Singer of Tales* (New York: Atheneum, 1974), source for the "Song of Bagdad." There are several attempts to retrieve fragments of Inca oral literature: all are too late; Christian and Western influences are painfully clear. For examples, see Inca selections in Sebastían Salazar Bondy, *Poesía Quechua* (Mexico, D. F.: Universidad Nacional Autónoma de Mexico, 1964); and John H. Rowe's analysis, *Eleven Inca Prayers from the Zithuwa Ritual,* Kroeber Anthropological Society Papers nos. 8, 9, (Berkeley, 1953), pp. 82–99.

The characterization of statistics as "political arithmetic" originated in seventeenth-century England. For more on statistics and its connection with states, see Harold Westergaard, *Contributions to the History of Statistics* (London: P. S. King and Son, Ltd., 1932). The connection continued into modern times as is clear in many of the statement connecting quipus and oral history. At least one thing can be concluded from what Cieza and others wrote: given the varied contexts in which quipus were used, they must have been the realization of a very general recording system.

It is easier to guess how the quipumaker might record a census, tribute, and the like, than it is to imagine how he might record oral history. There is no contradiction in the notion of a record of oral history: if history was sung, as in a ballad, the transmission was oral even if a record was made, just as the musical accompaniment for a ballad is not the ballad itself. But what do numbers have to do with oral history? Remember that numbers can be used to express quantities or as labels. In oral history, they were probably used both ways.

Used as labels, numbers could have served to mark a series of the same or closely related phases. Repetitions of a phrase, later followed by repetitions of a different phrase, or a return to an earlier phrase, are distinctive of the literature of peoples who transmit their tradition orally. Take, for example, a narrative from the native African state of Burundi. The narrative is about a king and his diviner who set out to overthrow another king. There are ten episodes, $A-J$. The first episode (A) is followed by three episodes (BCD). The starting phases of D repeat, with slight variation, the start of C; and the start of C is a repeat, with slight variation, of the starting phases of B. The story continues with a middle episode E; there is another set of repeats ($FGHI$) with the pattern shown below, and the narrative concludes with episode J. That is,

$$\underline{A} \; B \; C \; D \; \underline{E} \; F \; G \; H \; I \; \underline{J}.$$

Repetitions in oral literature are called formulas; they define the fundamental structure a narrative will take. The quipumaker could easily record the formulaic expression of the Burundi story. A quipu, where numbers served as labels only, could be designed in several ways. Here is a very simple possibility of our own making:

Fig. 4.3

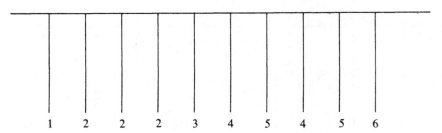

articles found in John Koren, ed., *The History of Statistics* (New York: Macmillan Co., 1918).

There are several general books on writing. I. J. Gelb, *A Study of Writing* (Chicago: University of Chicago Press, 1963), is comprehensive. The general issue of human communication, including writing, is treated in a book bearing that title: it is Colin Cherry, *On Human Communication,* (Cambridge, Mass.: MIT Press, 1966).

The Sumerian tablet cited in full is tablet 124 from James B. Nies, *Ur Dynasty Tablets* (Leipzig: Hinrichs'sche Buchhandlung, 1919). The general contents of tablets were compiled from several sources, including Edward Chiera, *Selected Temple Accounts from Telloh, Yokha and Dreham* (Philadelphia: University of Pennsylvania, 1922); and Tom B. Jones and John W. Snyder, *Sumerian Economic Texts from the Third Ur Dynasty,* (Minneapolis: University of Minnesota Press, 1961).

There is an assertion that writing antedates the Inca state in western South America. On this view see: R. Larco Hoyle, "La Escritura Peruana Sobre Pallares," *Revista de Geográfia Americana* 20 (1943): 345–54 and "La Escritura Mochica Sobre Pallares," ibid., 18 (1942): 93–103.

The oral world of the medieval scribe is described in H. J. Chaytor, "Reading and Writing," in *Explorations in Communication,* ed. E. Carpenter and Marshall McLuhan (Boston: Beacon Press, 1960). A highly informative essay on the world of the scribe in fourteenth-century England is "The English Civil Services in the Fourteenth Century," *The Collected Papers of Thomas Frederick Tout,* vol. 3 (Manchester: Manchester University Press, 1934). The consequences of the replacement of the acoustic by the visual in modern times, with particular regard to media, is the main concern in Marshall McLuhan, *The Gutenberg Galaxy* (New York: Mentor, 1969). The point is developed somewhat differently in K. Heyman and E. Carpenter, *They Became What They Beheld* (New York: Ballantine Books, 1970).

Another common feature of oral history is the incorporation of catalogs. There are four catalogs, for example, in the Slavic narrative "Song of Bagdad." They appear in the part of the narrative devoted to the summoning of chieftains to a meeting. The four catalogs are: invitation catalog (a1); ordering provisions (b1); arrival of provisions (b2); and arrival catalog (a2). Catalogs in oral history contain numbers that express quantities. It would be easy on a quipu to combine numbers that express quantities (such as those that occur in catalogs) with numbers that serve as labels (to record structures). One possibility is to place numbers at two levels: say, the lower ones for labels, the upper ones for quantities. For other aspects of state oral history, the quipumaker could use a variety of colors and cord spacings.

The paradigm for recording oral history fits peace negotiations, law, and calendrics. It also works for contexts mentioned nowhere in the Spanish chronicles in connection with quipus. For example, the quipumaker could make a recording of a completed building. He also could set down the plan for a building that was supposed to be erected and never was built. But there seems little question that he was most often doing statistics. Statistics originally meant the collection and processing of data in the interest of the state. In this sense, the one used here, statistics is political arithmetic. Image the quipumaker at work toward the end of a day in a storehouse area and you have a picture of him engaged in political arithmetic. Among other things, his quipu might show sums that were subtotals of different kinds of goods that left the storehouse on that day. One group of pendant cords (we do not know how many) along the main cord (of a length unknown to us) could look like this, where the pendant values (perhaps recordings of quantities of different kinds of food) are summed on a top cord:

Fig. 4.4

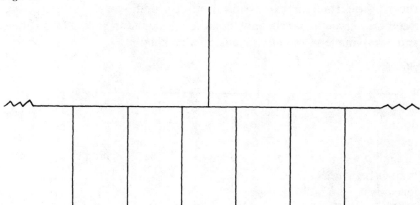

Can we settle the question of what the quipumaker knew by simply saying that he could write? Writing is a subject with several parts; the quipumaker wrote or did not depending on which part is taken into account. One part has to do with the notion of a general recording system. With every quipu, its maker realized a system that was general compared to special purpose systems, for example, records of chemical reactions. This should be clear from the preceding paragraphs. So, with regard to only this part of writing, the answer is that the quipumaker could write. Other parts of writing are its contents and setting. To examine these, we return to the Sumerian scribe.

The marks that the Sumerian pressed into clay were drawn from a repertoire of 600 or more signs. Of these, about 100 to 150 stood for syllables in his language. For example, one sign was used for the sound *ba*, another for *dal*. The Sumerians are credited as the first people to develop the use of speech sounds in a recording system. There are at least 100,000 known samples of this system, each one a separate clay tablet.

For the first thousand years of its existence, the system was used solely for political arithmetic. Scribes worked for the temple or for the palace. As already noted, his clay dried fast, so the scribe recorded daily transactions and later summarized a week, or a month, or a year's account on other tablets. These kinds of things were recorded: allowance for travel; account of a grain harvest; inventory of copper tools; rations of beer, bread, onions, oil, and spice. Here is the full content of a typical transaction: *Shirra received from Abbashagga 56 sheep, 6 lambs, 159 ewes, 99 female lambs, 5 milking goats, and 30 female kids, 6th month, Bur-Sin 2nd* (and in the margin, a total is given: 355). Another unexceptional tablet records the work done in eight palace workshops for a twelve month period. One side of still another tablet has already been shown (see plate 4.1). On this tablet, the scribe recorded payments, mostly in grain, to water pourers, guards, female singers, shepherds, and other workers in the temple of *Dumuzi*. The signs

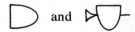

do not stand for syllables; they probably mean full time and half time, respectively.

Like the Sumerian scribe, the scribes in our culture use speech sounds in their recording system. Our culture calls what they do writing. Therefore—it is generally concluded—the Sumerian scribe, who for a thousand years used a similar system to write solely the political arithmetic described above, could write; others, who used systems that did not incorporate marks for speech, could not. Writing for the Sumerian scribe, however, was much different than it is for us. It is more than a matter of what is written and under whose auspices it is done. As children, we learn to write by saying words, hearing what we say, and making appropriate associations with what we see we are doing with our hands. By the time we are young adults, sound has been eliminated and replaced by vision to the point where we ask to see something—say a letter read to us—in order to grasp its meaning. Not so for the adult Sumerian scribe: he existed in a setting where hearing and saying words were never disconnected from seeing them, as it is with us, who write and read in silence. Some say that this is to our detriment. In any case, the difference is profound enough to define two distinct settings. In Western culture, the medieval scribe in his noisy booth was probably the last of his profession to understand recording in terms similar to the Sumerian scribe.

Now the quipumaker shared with the Sumerian scribe, the Egyptian, and indeed with all scribes in the formative periods of civilizations, these three things: (1) a general recording system; (2) contents, namely political arithmetic (they worked for the state); and (3) a setting where recording involved saying/hearing as well as seeing/touching. It seems fair to say that the quipumaker knew how to write. Yet, he did not write as we generally understand that term. He used cues from a shared informational model within Inca culture, and particularly those aspects of the model related to state affairs. A quipumaker told to make a recording of a village census brought to his task an understanding of what a census meant to the Incas. An Inca bureaucrat, sharing the same informational model and told that the quipu was the record of a village census, could read what the quipumaker recorded. We, for example, given something and told it is our calendar, would know to associate divisions of seven with days of the week in a certain order, and divisions into twelve with months in a certain order. And Americans, seeing a distinctive color or mark for the fourth day of the seventh month, would know that it means the memorial of the signing of the Declaration of Independence, while Mohammedans would recognize a special color for the twenty-seventh day of the ninth month as meaning the commemoration of Mohammed's receiving the Koran. However, we are not party to the informational model that would permit us to associate a particular quipu with a particular cultural meaning. For that reason, it is impossible to say that a certain quipu is a record of a certain census, or a song in praise of a Sapa Inca.

Study of quipus, however, can yield an understanding of the general recording system and some of the mathematical concepts, structural principles, and expressive techniques that underlie it. The following chapters convey our understanding of these. Your pleasure and understanding will be increased if you attempt to do the exercises we have provided for you.

Chapter 5

Format, Category, and Summation

1 Quipus are records. If many records, particularly those that contain numbers, were gathered together, they would differ from each other in form as well as in content. A monthly bank statement, a racing form, a ball game scorecard, a train schedule, a desk calendar, a stock market report, a telephone directory all look different but, as contrasted with most records, they share the special feature that each has a formal layout. Each has a format designed for the display of the data particular to it. A specific format has been arranged to convey the data concisely while displaying significant relationships and enabling significant comparisons. In fact, without the formality and consistency of the data arrangement, most of the information would not be accessible to the reader. Imagine, for a moment, a telephone directory in which either the given name or the surname preceded the other, in which sometimes the address came before the phone number and sometimes vice versa, in which no set spacing was used so there were no columns on the page, and in which names were not listed in alphabetical order. All the information would still be in the directory, but it is almost impossible to imagine finding anything.

Designing a format is an exercise in logic because its goal is an explicit visual statement of the logical structure of the information. Repetition is implicit in the formatting of information because it is categories within the data, not individual items or facts, that are being made explicit. In the example of a telephone directory, the format normally includes visual categorization into name, address, and phone number because these data are repeated for each individual. The use of alphabetically arranged names is a way to order the repetitious data.

Obviously, not all records contain formatted information. Also, if there were only one sample of a particular type of format, it might easily go unrecognized. Similarly, some quipus are formatted and some are not. Moreover, the fact that there are only a limited number of quipus still around means that some formats will not be recognized as such. However, the quipus that exist lead us to conclude that formatting was of very great importance to the quipumakers. In this chapter, we explore in detail the logical structure of one of the most prevalent quipu formats. It is characterized by cross categorization and summation. We will first build a description of cross categorization and then give some specific quipu examples. The ways that summation are involved will then be discussed and specifically illustrated.

2 When dealing with seven pieces of information that consist of, for example, weights of three people and outdoor temperatures at four times

of the day, the data can be described in two categories. They can be written in many different arrangements but, once having recognized that there are two categories, it would be more convenient for future use to record them as a group of three items and another group of four items. They can be written for example, as: w_1, w_2, w_3; t_1, t_2, t_3, t_4. Were these data to be recorded on a quipu, the cords could be arranged as a group of three pendants separated by a space along the main cord from a group of four pendants. Or, there could be three pendants of one color followed by four pendants of another color. To summarize these written or quipu arrangements, the mathematical notation w_i (i = 1, 2, 3), t_j (j = 1, 2, 3, 4) can be used. (This is read as: w sub i where i = 1, 2, or 3, and t sub j where j = 1, 2, 3, or 4). It simply means that two categories have been conveyed, one with three elements placed in order and the other with four elements in order.

Next, consider recording six pieces of information consisting of the scores of three games involving two teams. This data can be viewed as in two different categories: scores of Team 1 and scores of Team 2. Or, it can be viewed in three different categories because it describes three different games. Actually, both can be done simultaneously. The data can be organized into a chart:

Fig. 5.1

	Team 1	Team 2
Game 1		
Game 2		
Game 3		

Reading down the first column, the scores of Team 1 are associated, and reading down the second column, the scores of Team 2 are associated. Reading across the first row describes Game 1 while the second and third rows describe Games 2 and 3 respectively. Alternately, the chart used could be:

Fig. 5.2

	Game 1	Game 2	Game 3
Team 1			
Team 2			

Here columns associate a game's data while rows associate a team's data.

A quipu record of these data could use space and/or color to display the categories. One possibility is:

Fig. 5.3

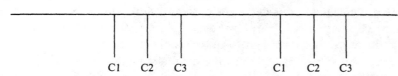

where pendants in the first group contain Team 1 scores while those in the second group contain Team 2 scores, and colors 1, 2, 3 are associated with Games 1, 2, 3 respectively.

Another possibility is:

Fig. 5.4

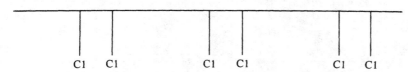

where all cords are the same color. The first group contains scores of Game 1, the second of Game 2, and the third of Game 3. Within each group, the first position is associated with Team 1 and the second position with Team 2.

In subscript notation, both charts and both quipus are summarized by p_{ij} ($i = 1, 2, 3$; $j = 1, 2$). (Read as: p sub ij where $i = 1, 2,$ or 3 and $j = 1$ or 2.) A value of i is associated with each game and a value of j is associated with each team. Thus p_{11} refers to the score in Game 1 of Team 1 and p_{32} refers to the score in Game 3 of Team 2. To further clarify the meaning of the subscripts, figure 5.5 combines the chart forms and the quipu forms with the subscript form.

Fig. 5.5.

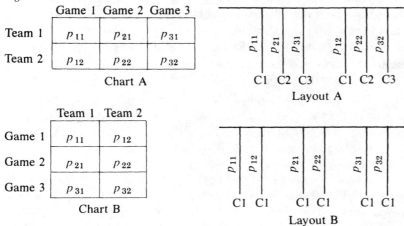

Before proceeding to a generalization, another step of cross categorization is needed. Consider a set of twenty-four pieces of information which are the prices of three products in four stores on two different days. To write them, we can form one table for each day as shown in figure 5.6. In subscript notation, the data is summarized as p_{ijk} ($i = 1$, 2; $j = 1, 2, 3, 4$; $k = 1, 2, 3$) where i is associated with day; j is associated with store; and k with product. The twenty-four individual p's have been placed in the tables in figure 5.6 in order to show how they correspond to the tabular form.

Fig. 5.6

Day 1

	Prod. 1	Prod. 2	Prod. 3
Store 1	p_{111}	p_{112}	p_{113}
Store 2	p_{121}	p_{122}	p_{123}
Store 3	p_{131}	p_{132}	p_{133}
Store 4	p_{141}	p_{142}	p_{143}

Day 2

	Prod. 1	Prod. 2	Prod. 3
Store 1	p_{211}	p_{212}	p_{213}
Store 2	p_{221}	p_{222}	p_{223}
Store 3	p_{231}	p_{232}	p_{233}
Store 4	p_{241}	p_{242}	p_{243}

There are several other ways the data could be arranged in tabular form. The two tables of figure 5.6 could, instead, have rows for products and columns for stores; there could be three tables of eight entries each, where each table is associated with a product (and either rows or columns with days and stores); and there could be four tables of six entries each, where each table is associated with a store (and either rows or columns with days and products). No matter which of these tabular formats is used, the same summary subscript notation applies because each format is a way of demonstrating the same cross categorization. Thus, the subscript notation is identifying the logical structure being exhibited by them regardless of which of the ways it is exhibited. It is for this reason that the subscript notation has been introduced and is preferable. When describing the logical structure of actual quipus, the subscript notation enables a concise exhibition of the structure without having to include the particular details of the quipu construction. Moreover, a statement of the logical structure provides the framework for examining the data within it. Without an understanding of the logical organization of a telephone directory, looking for someone's number would be as hopeless a task as looking in a directory that had no logical organization.

For the same data as in figure 5.6, many different quipu arrangements could be used. A few of them are shown in figure 5.7. Each has twenty-four pendants and conveys by spacing and color that four categories are being crossed with three categories which are being crossed with two categories. All are examples of the summarizing subscript notation p_{ijk} ($i = 1, 2$; $j = 1, 2, 3, 4$; $k = 1, 2, 3$).

Fig. 5.7.

C1C1C1C1 C1C1C1C1   C1C1C1C1C1C1C1C1   C1C1C1C1C1C1C1C1

groups (by large spaces) correspond to	$k = 1, 2, 3$	(product)
subgroups (by smaller spaces) correspond to	$j = 1, 2, 3, 4$	(store)
position (in subgroups) correspond to	$i = 1, 2$	(day)

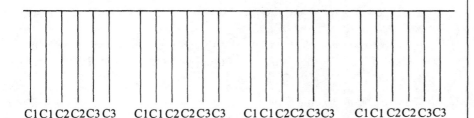

C1C1 C2C2C3 C3   C1C1 C2C2C3C3   C1 C1 C2C2 C3C3   C1C1 C2C2 C3C3

groups (by large spaces) correspond to	$j = 1, 2, 3, 4$	(store)
subgroups (united by same color) correspond to	$k = 1, 2, 3$	(product)
position (in subgroups) correspond to	$i = 1, 2$	(day)

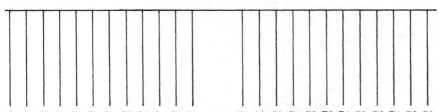

C1C2 C3C1 C2 C3 C1 C2 C3 C1 C2 C3   C1 C2 C3 C1 C2 C3 C1 C2 C3 C1 C2 C3

groups (by large spaces) correspond to	$i = 1, 2$	(day)
subgroups (united by color pattern) correspond to	$j = 1, 2, 3, 4$	(store)
position (in subgroups) correspond to	$k = 1, 2, 3$	(product)

3 The quipus in figure 5.8 are:
1 (AS74); 2 (AS66); 3 (AS84);
4 (AS114); 5 (AS57); 6 (AS32);
7 (AS175); 8 (AS121).

3 Many quipus are formatted so that they display cross categorizations. Primarily, it is spacing and color coding that convey categories. Sometimes, for emphasis or as an extension of these devices, additional visual markers are used. These markers include very long cords, very short cords, cord puffs tied on the main cord, or specially colored cords. A sample of the larger quipus of this type are summarized in figure 5.8. Some important ideas to keep in mind when looking at them:

1. The logical complexity of cross categorization increases as the number of subscripts increase. Our example used first two subscripts and then three subscripts. Two subscripts are analogous to a table of data and three subscripts to a set of several tables of the same size. An analogy to four subscripts could be several pages each containing several tables.

2. The highest value for each subscript is the number of logical distinctions being made. Our examples had highest values of 2 and 3 and of 2, 3, and 4. These values are analogous to the number of rows or columns in a table and then to the number of tables and the number of pages of tables.

3. The product of the highest value for all subscripts gives the number of items in the entire set of data. Our examples had $2 \times 3 = 6$ items and $2 \times 3 \times 4 = 24$ items.

Fig. 5.8. Some quipu configurations.

1. p_{ij} $i=1, \ldots, 5$; $j=1, \ldots, 17$

2. p_{ij} $i=1, \ldots, 8$; $j=1, \ldots, 24$

3. p_{ij} $i=1, \ldots, 15$; $j=1, \ldots, 25$

4. p_{ijk} $i=1, 2$; $j=1, \ldots, 10$; $k=1, \ldots, 16$

5. p_{ijk} $i=1, \ldots, 4$; $j=1, \ldots, 6$; $k=1, \ldots, 24$

6. p_{ijk} $i=1, \ldots, 6$; $j=1, \ldots, 8$; $k=1, \ldots, 9$

7. p_{ijkm} $i=1, 2$; $j=1, 2, 3$; $k=1, \ldots, 5$; $m=1, \ldots, 7$

8. p_{ijkm} $i=1, 2$; $j=1, 2, 3$; $k=1, \ldots, 7$; $m=1, \ldots, 10$

4 To introduce summation, we return to the example of two teams competing against each other in three games. Now the record (shown in fig. 5.5) is also to include the sums of points scored per game and the sums of points scored per team.

Fig. 5.9. Chart B of figure 5.5 extended.

	Team 1	Team 2	Game Sums
Game 1	p_{11}	p_{12}	$p_{11}+p_{12}$
Game 2	p_{21}	p_{22}	$p_{21}+p_{22}$
Game 3	p_{31}	p_{32}	$p_{31}+p_{32}$
Team Sums	$p_{11}+p_{21}+p_{31}$	$p_{12}+p_{22}+p_{32}$	

Summation of the scores in the first row of chart B results in the total points in Game 1 and summation of the second and third rows result in the total points in Games 2 and 3 respectively. These are recorded in a column that has been appended to the chart in the row associated with the game being described (see fig. 5.9). Similarly, the summation of the first column results in the total scored by Team 1, and summation of the second column gives the total for Team 2. In order to associate each of these totals with the appropriate team, they are placed in the team's column in an appended row. Whatever format was originally designed to record the scores can be enlarged to include the sums in keeping with their associated categories.

If the data had been recorded on a quipu, the placement of the sums would also depend on how the categories had been designated. Where the quipu had two spatial groups to distinguish the scores by team (see fig. 5.5, layout A), the sum for each team would remain with the group. It could be on a top cord associated with the group or on an additional cord in the group (fig. 5.10).

Fig. 5.10. Layout A of figure 5.5 extended in two possible ways.

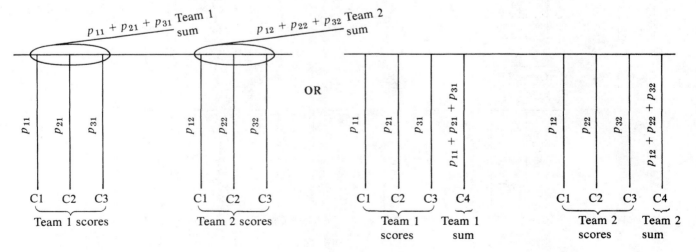

Since, within each group, the scores for different games were distinguished by color, the sums for each game would be in another group still identified by their color (fig. 5.11).

Fig. 5.11. Layout A of figure 5.5 extended.

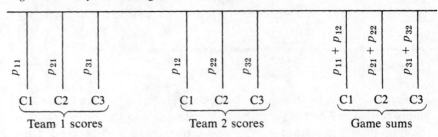

Another possible quipu layout was shown in figure 5.5 (layout B) for the same data. On it, game scores were distinguished by spatial groups while team scores were distinguished by pendant position within a group. Therefore, in this layout, sums of scores would be placed and identified in kind. Figure 5.12 shows layout B extended to include sums. Although shown separately, both game and team sums could be included on the same quipu.

Fig. 5.12. Layout B of figure 5.5 extended.

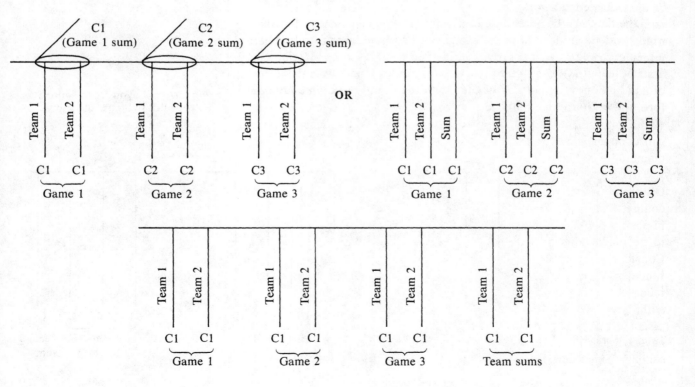

If the three total game scores were summed, the result would be the same as if the two total team scores were summed since each would actually be the grand total of the six individual scores. This value is, therefore, associated with the entire quipu regardless of its particular layout. If present, it could be on a first or last pendant set off from the rest, on a dangle end cord, or even knotted into the main cord itself.

5 Cross categorization with categorical summation is present on about 25 percent of the quipus. Depending on the format used for categorization, the sums are on top cords associated with groups, within associated groups, or on pendants in groups that sum other groups. Grand totals, which unite all categories, are separately and prominently placed.

Several quipus have more elaborate layouts involving summation. Before proceeding to them, two observations fundamental to all summation need to be made explicit. In our earlier discussion of numbers, it was noted that not all numbers are magnitudes. Numbers that are labels, such as apartment numbers in a multifloor building, are not summable. The numbers in exercise 2.1 were the opening and closing hours of a service station so their sum would have no meaning. Even when numbers are magnitudes, they might not be additive. If six pieces of data were the heights and weights of three individuals, recording the sum of the two values associated with each individual would be highly unlikely. The sum of the three items in the weight category could be of interest. For example, contemporary passenger elevators usually carry a statement of maximum total weight permitted. It is difficult to think of a reason for summing the three items in the height category. Therefore, the presence of summation is a significant statement about the rest of the data on the quipu. When a quipu contains categorical sums, we can conclude that the data being summed are magnitudes and that the magnitudes are such that their sum is meaningful.

The second observation relates to the number zero and color coding. On some quipus, cord positions within a group are reinforced by color coding. That is, when, for example, colors 1, 2, 3, 4 appear in the groups in positions 1, 2, 3, 4 respectively. This redundancy enables a distinction to be made between a value of zero and the absence of a category. Consider a set of data that is an inventory of the number of pairs of sandals owned by a community of people. Say that the numbers are recorded on pendants categorized into families by spatial groups and, within each group, categorized into men, women, and children as represented by cord colors 1, 2, 3. A cord of color 3 with no knots would have a value of zero and mean that the children of a family had no sandals. The absence of cord color 3 in the cord group would mean that

5 On fourteen of the eighteen quipus that have top cords, the top cords carry the sums of the associated groups. Two more probably would also but the knotting of numbers onto cords seems incomplete. An interesting sidelight is that on twelve of these eighteen quipus, the pendant groups are of size six. In the general sample, of those quipus with distinct pendant groups, only about 25 percent have groups of size six. Therefore, pendant groups of size six are more prominent (67 percent) among pendant groups with top cords than among pendant groups in general (25 percent). Sums appear within their associated groups on AS98, AS152, and AS159. Sums are on pendants in groups that sum other groups on AL4, AL7, AS8, AS49, AS100, AS120, AS143, AS147, AS176, and AS178. Grand totals are on L49, AS100, AS122, AS127, AS130, and AS178.

the family had no children. These are different in information content but either contributes nothing to a categorical sum. The fact that the distinction is made is important because it corroborates that a blank cord, in fact, represents what is symbolized by our zero. It also provides the key to many quipu formats. In cases where there are spatial groups with different numbers of cords but all are subsets of one color pattern, the logical structure can be identified from this overall color pattern.

6 To further elaborate the cross categorization and summation, we again return to the score chart of the two teams competing in three games. The complexity of the chart is increased by including the fact that each team has four players and their individual scores are also to be included. Figure 5.13 is the extended chart but still includes all the data and sums from figure 5.9. The p_{ijk} ($i = 1, 2, 3; j = 1, 2; k = 1, 2, 3, 4$) representing the scores of each player k on team j in game i, have also been noted on the chart. It is important to realize that the team scores for each game are now sums of player scores. Therefore, each team total is the sum of three game scores which in turn are each sums of four player scores. Similarly, the three game totals are each sums of pairs of team scores which in turn are sums of player scores.

Fig. 5.13. **Figure 5.9 extended.**

	Player 1	Player 2	Player 3	Player 4	Team 1
Game 1	p_{111}	p_{112}	p_{113}	p_{114}	p_{11}
Game 2	p_{211}	p_{212}	p_{213}	p_{214}	p_{21}
Game 3	p_{311}	p_{312}	p_{313}	p_{314}	p_{31}
					$p_{11}+p_{21}+p_{31}$ Team Sum

	Player 1	Player 2	Player 3	Player 4	Team 2
Game 1	p_{121}	p_{122}	p_{123}	p_{124}	p_{12}
Game 2	p_{221}	p_{222}	p_{223}	p_{224}	p_{22}
Game 3	p_{321}	p_{322}	p_{323}	p_{324}	p_{32}
					$p_{12}+p_{22}+p_{32}$ Team Sum

Game Sums

$p_{11}+p_{12}$
$p_{21}+p_{22}$
$p_{31}+p_{32}$
$p_{11}+p_{12}+p_{21}+p_{22}+p_{31}+p_{32}$ Grand Total

This type of cross categorization and summation can be viewed in terms of other familiar contemporary situations. An example is an accounting scheme for a company with several departments. A final report might include total expenses categorized by department. For each department there is a ledger elaborating the different types of expenses; say sales expense, equipment expense, expenses of utilities and maintenance, and salaries. Then, for each department for each of these expense categories, there is a subledger further detailing the expenses. Another example is a company that makes several products and organizes its sales into geographic regions. The final report would be the total number of each kind of product sold, the charts would be products sold categorized into sales regions, and each sales region might have a subchart further detailing its distribution of each product. In these examples, as contrasted with the team scores, the subledgers (or subcharts) do not necessarily have the same number of categories within them and some ledgers might not even have associated subledgers.

To summarize the examples in terminology that will be used for the quipu descriptions, we have described:

   1. Team scores per game—these form a chart of p_{ij} ($i = 1, 2, 3$; $j = 1, 2$)—each p_{ij} is the sum of four items in an associated subchart.
   2. Player scores p_{ijk} ($i = 1, 2, 3$; $j = 1, 2$; $k = 1, 2, 3, 4$) constituting two subcharts, one for each value of j.
   3. Categorical sums: team sums ($j = 1, 2$); game sums ($i = 1, 2, 3$).
   4. Grand total.

7 On several elaborate quipus, some of these arrangements are found. Here are the details of four quipus involving subcharts.

7 Quipu example 5.1 is AS156. Quipu example 5.2 is AS38 (the quipu shown in plate 2.6). Quipu example 5.3 is AS149. Quipu example 5.4 is AS175.

Quipu Example 5.1

   1. Chart with elements p_{ij} ($i = 1, 2, 3, 4$; $j = 1, 2$)—each p_{ij} is the sum of seven items in an associated subchart.
   2. Elements p_{ijk} ($i = 1, 2, 3, 4$; $j = 1, 2$; $k = 1, \ldots, 7$) constituting two subcharts, one for each value of j.
   3. This quipu, therefore, has sixty-four values. There are sixty-four pendant cords suspended from the main cord. By a repeated color pattern and spacing, these are separated into sixteen groups of four pendants each. By spacing and a marker, the sixteen groups are separated into two, seven, seven. The actual data on the pendant cords is shown in tabular form (fig. 5.14). In the table, each group of

four values is in a row. Consecutive groups are placed in rows beneath each other rather than strung out horizontally as they are on the quipu.

Fig. 5.14. Quipu example 5.1.

		$i=1$	$i=2$	$i=3$	$i=4$
Chart	$j=1$	160	350	160	330
	$j=2$	140	360	140	360
	$k=1$	20	50	20	50
	$k=2$	20	40	20	40
	$k=3$	20	50	20	50
$j=1$ Subchart	$k=4$	40	60	40	60
	$k=5$	20	50	20	50
	$k=6$	20	50	20	40
	$k=7$	20	50	20	40
	$k=1$	20	52	20	51
	$k=2$	20	52	20	51
	$k=3$	20	52	20	51
$j=2$ Subchart	$k=4$	20	52	20	52
	$k=5$	20	52	20	51
	$k=6$	20	50	20	52
	$k=7$	20	50	20	52

Quipu Example 5.2

1. Chart with elements p_{ij} ($i = 1, 2; j = 1, 2, \ldots, 18$)—each p_{ij} for which $i = 1$ is the sum of three items in an associated subchart.

2. Elements p_{1jk} ($j = 1, 2, \ldots, 18; k = 1, 2, 3$) constituting one subchart for $i = 1$.

3. Categorical sums ($j = 1, 2, \ldots, 18$).

Quipu Example 5.3

1. Chart with elements p_{ij} ($i = 1, 2; j = 1, 2, 3, 4, 5$)—each p_{ij} is the sum of seven items in an associated subchart.

2. Elements p_{ijk} ($i = 1, 2; j = 1, 2, 3, 4, 5; k = 1, 2, \ldots, 7$) constituting two subcharts, one for each value of i. The five elements for which $i = 1$ and $k = 1$ are each the sums of nine elements in an associated sub-subchart.

3. Elements p_{1j1m} ($j = 1, \ldots, 5; m = 1, 2, \ldots, 9$) constituting one sub-subchart for $i = 1, k = 1$.

Quipu Example 5.4

1. Chart A with elements p_{ij} ($i = 1, 2, 3; j = 1, 2, 3, 4, 5$)—each p_{ij} is the sum of seven elements in an associated subchart.

2. Elements p_{ijk} ($i = 1, 2, 3; j = 1, \ldots, 5; k = 1, \ldots, 7$) constituting three subcharts, one for each value of i.

3. The quipu contains another set of subcharts (B) with elements q_{ijk} ($i = 1, 2, 3; j = 1, \ldots, 5; k = 1, \ldots, 7$). The main cord of the quipu is evidently broken. Because of the layout of the quipu,

broken end—subcharts B—chart A—subcharts A

we believe that it contained another portion that would have been chart B with elements q_{ij} ($i = 1, 2, 3; j = 1, 2, 3, 4, 5$).

We have no way of knowing where the data on these quipus came from. But we can recognize formats, and indeed, the specifics of the format. This, in itself, tells us something very important about the Incas, namely the formal specific and rather complex ways in which they ordered their universe.

8 Before concluding the discussion of cross categorization and summation, the word *sum* needs some further examination. When we see the values 18, 31, and 49, we can describe 49 as the sum of 18 and 31. However, this carries no implication that we have seen the *operation* of

addition carried out. A quipu is not a calculating device. We do not know how or where the arithmetic was done. In fact, we do not even know what arithmetic was done. Just a few examples of how these numbers could be operationally related are:

18 results when 31 is subtracted from 49;
18 added to 31 results in 49;
49 can be subdivided into 18 and 31.

If there were only these three values, the arrangement might be fortuitous and we would even hesitate to presume there was some arithmetic relationship. Because the logical structure of the data was identified first, and because, within that structure, there is repetition of a patterned relationship, we can proceed to state the pattern. This still leaves as an open question the human actions that placed the data into this relationship.

9 Charts and subcharts of ball game scores might be a prediction for the next season rather than the results of games played. A budget which allocates money that is available to be spent next year usually has the same classificatory scheme as a report of expenditures for last year. And a record of number of products by sales regions could represent a plan or goal rather than a report of past sales. The important difference is that, in these latter cases, the summary numbers are determined first and then broken down into parts.

The data within the charts and subcharts of some of the quipus have a regularity that is more suggestive of a plan than of a result. Or, if it is a result, it is the result of something that was very carefully planned. Regularity of data is hard to define because it can show up in many different ways: values that are exactly the same (200, 200); values that are multiples of a common unit (155, 85, 315, 500); values that are a fixed proportion of each other (41, 82; 71, 142; 193, 386); values that are symmetrically placed (48, 11, 91, 11, 48). When, as in one of the quipu examples below, ten values show regularity but each is the sum of seven other values and some of these are, in turn, the sum of nine other values, it is more plausible that the ten values were determined first and then broken down into the associated 105 numbers than that the ten values were built up from them.

In order to find regularities, the actual numbers on a quipu have to be examined. We first looked at the logical structure of the quipus and then at the arithmetic relationships within them. Summations, charts, and subcharts are common to both a record of results and a plan. Keeping within this common framework, but delving into the actual numbers

9 Quipu example 5.3—*continued*:
To clarify the effect that a large range of values has on comparisons, consider the following situation. The prices of $1.92 and $1.94 for an item are not the same. But when faced with prices of $0.06, $1.92, $1.94, and $84.72, the two would appear much more similar. Having only integers also effects comparisons. If fractions were available, we would not say that 1, 3, 5, 6 are each half of 2, 7, 10, 12 because half of 7 is 3 1/2. With no fractions available, regarding 3 or 4 as half of 7 becomes appropriate. These notions are discussed further in chapter 7.

within the relationships, we can attempt to distinguish between them. Our question is whether the sums were formed from the values in the subcharts or whether the sums were preconceived values that were then decomposed.

Two of the four quipus that were just used as examples (quipu examples 5.1 and 5.3), exhibit distinctive data regularities. We, therefore, return to look at each of them.

Quipu Example 5.1—*Continued*
All the values on the quipu were shown in Figure 5.14.

1. The eight chart values are each the sum of seven subchart values. But when the chart's categorical sums are formed (by adding together the four values for which $j = 1$ and by adding together the four values for which $j = 2$), each turns out to equal 1,000.

2. Within each of the subcharts there are twenty-eight values. However, each has only four different numbers used over and over again. When the twenty-eight values are combined into fourteen pairs (for each k, values for $i = 1$ and $i = 2$ are combined; values for $i = 3$ and $i = 4$ are combined), the regularities within each subchart and between the subcharts become more apparent. Each subchart is rewritten as a table of pairs in figure 5.15.

Fig. 5.15. Quipu example 5.1—*continued*

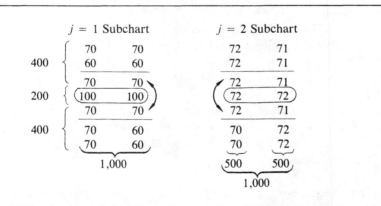

a) Each pair table (subchart) has only three different numbers. One has only multiples of 10 and the other has consecutive numbers. In each case, there are 8, 4, and 2 of each of its three numbers.

b) In each, the center row has a pair of maximum values.

c) In each, the rows just above and below the center row are the same.

d) In the $j = 1$ subchart, there is symmetry of the sum of all values above and below the center row: 400, 200, 400.

e) In the $j = 2$ subchart, there is symmetry of the sum across a center line (vertically): 500, 500.

Quipu Example 5.3—*Continued*

    1. The chart of ten values is related to subcharts and sub-subcharts containing, in all, about a hundred values. However, when the five values in each of the two categories of the chart are compared, they have a remarkable numerical regularity. They are found to represent the same proportions of their own category. Because the numbers in the chart range from 660 to 21,243 and the proportions are not simple fractions, this might not be visible by looking at the ten numbers. To make the similarity apparent, each value has been divided by the sum of its i category so that only the proportions themselves are written in a table in figure 5.16.

Fig. 5.16.   Quipu example 5.3.

	$j = 1$	$j = 2$	$j = 3$	$j = 4$	$j = 5$
$i = 1$	0.121	0.017	0.535	0.105	0.221
$i = 2$	0.122	0.017	0.533	0.105	0.223

As usual, when dealing with real data, we do not expect the numerical exactness that is found in made-up textbook problems. Also, since all the numbers on the quipu are integers, with no fractions or decimal fractions, accuracy of individual numbers is limited to the closest whole integer. The numbers in the table in figure 5.16 are, therefore, rounded to three places and we accept as "the same" values that differ by at most 1 percent.

    2. In addition to the proportions being the same, one of the proportions is the sum of two others. That is, the proportion of the wholes for $j = 2$ and $j = 4$ combined, is the same as the proportion for $j = 1$.

    3. Another regularity on this quipu is in the data in subcharts and sub-subcharts. Many of the values are multiples of the value 110. In fact, in one of the subcharts, 38 percent of the entries are multiples of this single value.

10 There is another type of quipu, similar in concept to the kind being discussed. They show cross categorization and have internal regularities but do not have *on them* categorical sums. This lack is significant. Because they lack this evidence of summation, it cannot be presumed that the values are magnitudes which are additive. It is only when we form the sums and find that they are patterned, that we believe the interpretation is valid. These then are charts or subcharts. What makes them

10 Quipu example 5.5 is AS168. Quipu example 5.6 is AS141. Quipu example 5.7 is AS136. Its companion is AS140. Other quipus suspended from wooden bars are in the same museum. They are AS106, AS112, and AS124. Max Schmidt, *Kunst und Kultur von Peru* (Berlin: Impropyläen-Verlagzu, 1929), contains photos (p. 545) and a brief description (p. 97) of AS136 and its companion AS140. Quipu example 5.8 is AS109.

identifiable as such is their logical structure combined with the regularity of categorical sums *formed by us*. Two small examples of these follow.

Quipu Example 5.5
1. Chart with elements p_{ij} ($i = 1, 2; j = 1, \ldots, 9$).
2. Equal categorical sums for $i = 1$ and $i = 2$. (Nine values go into each sum.)

Quipu Example 5.6
1. Chart with elements p_{ij} ($i = 1, 2; j = 1, \ldots, 10$).
2. Equal categorical sums for $i = 1$ and $i = 2$. (Ten values go into each sum.)
3. The twenty values in the chart are only four different numbers repeated over and over.
4. In the chart, there is symmetry of the sum of all values above and below a center line and across a center line (fig. 5.17).

Fig. 5.17. Quipu example 5.6.

	$i = 1$	$i = 2$
$j = 1$–5	110	110
$j = 6$–10	110	110

Two larger examples are of considerable interest: one because of its patterning and the other because of an unusual construction feature.

Quipu Example 5.7
1. This quipu is distinctive in its construction because it is strung from a carved wooden bar (see plates 5.1 and 5.2). Unfortunately, the carvings do not help us to identify what is being recorded. It consists of two subcharts with elements p_{ijk} ($i = 1, 2; j = 1, \ldots, 9; k = 1, \ldots, 10$). Because of the way the quipu is attached to the bar, the subcharts are physically back-to-back. It is evidently related to another quipu believed to come from the same set. The related quipu is also strung from a carved wooden bar and uses the same style of color coding (see plates 5.3 and 5.4). It too, has a logical structure that can be described as q_{ijk} ($k = 1, 2; j = 1, \ldots, 9; k = 1, \ldots, 10$).
2. The subcharts of example 5.7 and their regularities are diagrammed in figure 5.18.

Plate 5.1. Quipu example 5.7. *(Courtesy of the Museum für Völkerkunde, Berlin, Germany.)*

CODE OF THE QUIPU

Plate 5.2. Carving detail from plate 5.1. *(Courtesy of the Museum für Völkerkunde, Berlin, Germany.)*

Plate 5.3. Companion to quipu **example** 5.7. *(Courtesy of the Museum für Völkerkunde, Berlin, Germany.)*

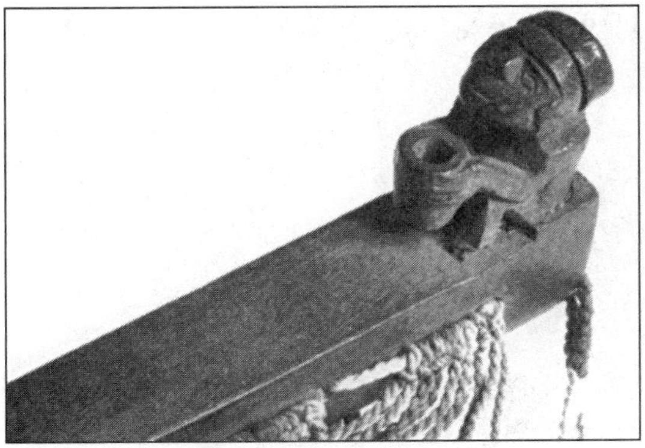

Plate 5.4. Carving detail from plate 5.3. *(Courtesy of the Museum für Völkerkunde, Berlin, Germany.)*

Fig. 5.18. Quipu example 5.7.

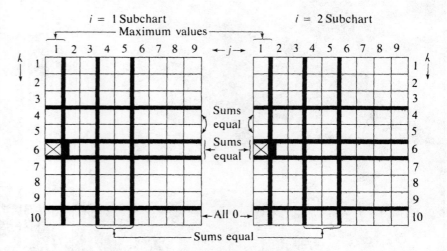

a) Category $k = 10$ contains all zeroes in both subcharts.
b) The categorical sum for $k = 6$ (excluding the value for $j = 1$) is the same value in both subcharts.
c) In each subchart, the categorical sum for $k = 4$ equals the categorical sum for $k = 5$.
d) The categorical sum for $j = 4$ and $j = 5$ combined, is the same in both subcharts.
e) In both subcharts, for each k, the values associated with $j = 1$ are greater than those associated with $j = 2, \ldots, 9$.
f) In addition to the relationships in the diagram: the categorical sum for $j = 3$ on $i = 1$ subchart equals the categorical sum for $k = 4$ on $i = 2$ subchart; and the categorical sum for $j = 2$ and $j = 3$ on $i = 1$ subchart equals the categorical sum for $j = 6$ and $j = 7$ combined on $i = 2$ subchart.

Quipu Example 5.8

1. Seven subcharts labeled $i = 1, 2, \ldots, 7$. They are of different sizes. By color coding, the smaller subcharts are related to parts of the larger ones. Notice in the description of the seventh subchart that there are only the second, third, and fourth categories of those identified with j.

a) For $i = 1, 3, 4$ p_{ijk} $(j = 1, \ldots, 5; k = 1, \ldots, 10)$;
b) For $i = 2$ p_{ijk} $(j = 1, \ldots, 5; k = 1, \ldots, 11)$;
c) For $i = 5$ p_{ijk} $(j = 1, 2, 3; k = 1, \ldots, 10)$;
d) For $i = 6$ p_{ijk} $(j = 1, \ldots, 5; k = 1, \ldots, 8)$;
e) For $i = 7$ p_{ijk} $(j = 2, 3, 4; k = 1, \ldots, 10)$.

2. We calculated the categorical sums which form the chart p_{ij} $(i = 1, \ldots, 7; j = 1, \ldots, 5)$. Each p_{ij} is the sum of eight, ten or eleven values depending on how many categories were identified with k. To display the regularity of these sums, they are written in tabular form in figure 5.19.

Fig. 5.19. Quipu example 5.8 sum table.

	$j = 1$	$j = 2$	$j = 3$	$j = 4$	$j = 5$
$i = 1$	766	1,572	1,120	640	1,007
$i = 2$	755	1,501	1,100	330	1,126
$i = 3$	755	1,501	1,100	330	1,136
$i = 4$	755	1,501	1,100	330	1,136
$i = 5$	467 ⎫ 1,000	1,501	1,100	—	—
$i = 6$	533 ⎭	1,200	1,052	112	1,120
$i = 7$	—	1,514	0	0	—

3. Within each of the subcharts there is repetition of groups of values. The repeated values differ from subchart to subchart.

a) In subchart $i = 1$, one grouping of five values (associated with $j = 1, 2, \ldots, 5$) is repeated twice ($k = 2, 3$) and another is repeated three times ($k = 6, 7, 9$).

b) In subchart $i = 2$, one grouping of five values (associated with $j = 1, \ldots, 5$) is repeated three times ($k = 3, 6, 9$).

c) In subchart $i = 3$, one grouping of five values (associated with $j = 1, \ldots, 5$) is repeated three times ($k = 2, 3, 5$).

d) In subchart $i = 6$, two groupings of four values (associated with $j = 1, 2, 3, 5$) are each repeated twice ($k = 2, 4; k = 6, 7$).

e) In subchart $i = 7$, one grouping of three values (associated with $j = 2, 3, 4$) is repeated seven times ($k = 1, 2, 3, 4, 5, 6, 7$).

For this quipu, particularly from the sum table (fig. 5.19), we invite the reader to find as many regularities as he can. Different readers may find different ones. But, in any case, we think they demonstrate that the logical structure of the quipu has been correctly identified from the cord spacing and color coding; that the data is additive and so the process of summation justified; and that the data is reflecting something planned carefully and with attention to detail.

11 The next chapter will continue with other logical structures. A later chapter will deal with other evidence of numerical and arithmetic relationships. But the particular quipus described earlier—those with cross categorization and summation—compactly characterize Inca insistence. There is formality of structure, both physical and logical. Cord placement, color coding, and number representation are the basic construction features repeated and recombined to define a format and con-

vey a logical structure. The structure itself is based on repetition and recombination of smaller units into larger units. There are categories grouped into subcharts and charts with sums and sums of sums. The data within the structure shows repetition and recombination as values are repeated in groups as well as summing to repeated or symmetrical values. These quipus certainly were constructed after careful planning but many also are part of a planning process.

EXERCISES

Exercise 5.1
Three sheds are being built. They are different sizes but each has walls made of cinder block and a floor and roof made of wooden boards. The materials used for the first shed are: 284 cinder blocks, 100 pounds of mortar, 28 boards, and 200 pounds of nails. For the second and third sheds respectively the materials are: 244 cinder blocks, 85 pounds of mortar, 24 boards, 170 pounds of nails; and 364 cinder blocks, 150 pounds of mortar, 51 boards, and 400 pounds of nails.

Design a quipu and record on it the amount of each material used for each shed and their sums. Show the quipu on a schematic including relative cord placement, cord color, knot types, and relative knot placement.

An Answer to Exercise 5.1

Exercises Quipu example 5.9 in exercise 5.3 is AS199. For purposes of the exercise we have modified one digit by 1 and another digit by 2. Original errors in counting, errors in arithmetic, errors in knotting, and our errors in counting knots or in transcription all are possible. Therefore, we do not consider a few errors of 1 or 2 in individual knot positions to be very significant.

Fig. 5.20

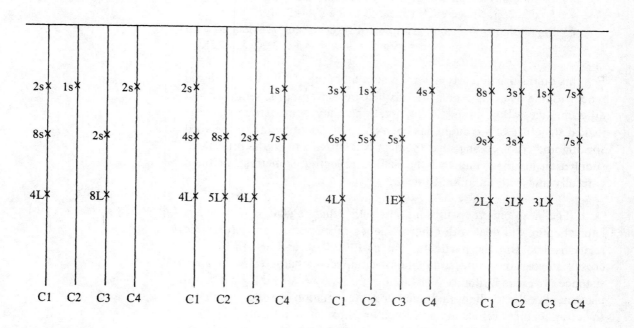

C1, C2, C3, C4 are associated with cinder blocks, mortar, boards, and nails respectively. Cord groups 1, 2, 3 are associated with sheds 1, 2, 3 respectively. Cord group 4 is associated with the material totals. (Note that there are no sums associated with the individual sheds as the amounts of four different materials are not additive.)

Exercise 5.2

A small community consists of four families. The families have the following members:

> Family 1: man, woman, man's mother, man's father, two children
> Family 2: man, woman, woman's older sister, four children
> Family 3: man, woman
> Family 4: man, woman, three children

One hundred potatoes and 242 ears of maize are to be distributed among the families. Each person in the community is to get the same share. However, no potatoes or ears of maize are to be chopped in pieces. The distribution among families is to be carried out with whole items. Any excess is to be divided between the two largest families who will distribute it to the oldest members of their household.

1. Calculate the number of potatoes and the number of ears of maize to be distributed to each family. Write the results in a table.

2. For each family, calculate the number of potatoes to be distributed to adults and the number to be distributed to children. Do the same for the ears of maize. Write the results in a pair of tables.

3. Design a quipu and record on it the distribution plan as described by the answers to parts 1 and 2. Give the answer in the form of a schematic.

An Answer to Exercise 5.2

1. There are twenty people in the community. Since there are 100 potatoes and each person is to get the same share, each person receives 5 potatoes. Family 1 has six people so gets 30 potatoes, family 2 with seven people gets 35 potatoes, family 3 with two people gets 10 potatoes, and family 4 with five people gets 25 potatoes.

There are 242 ears of maize to be divided equally among twenty people. Each person's share is 12. However, this leaves 2 ears of maize. The two largest families are families 1 and 2 so each of them gets 1 extra ear of maize. Family 1 with six people gets $6 \times 12 + 1$ ears of maize, family 2 with seven people gets $7 \times 12 + 1$ ears of maize, family 3 gets 2×12 ears, and family 4 gets 5×12 ears.

Fig. 5.21

	Fam. 1	Fam. 2	Fam. 3	Fam. 4
Potatoes	30	35	10	25
Maize	73	85	24	60

Fig. 5.22

2.

	Fam. 1	Fam. 2	Fam. 3	Fam. 4		Fam. 1	Fam. 2	Fam. 3	Fam. 4
Adults	20	15	10	10	Adults	49	37	24	24
Children	10	20	—	15	Children	24	48	—	36
		Potatoes					Maize		

Here are some observations on the data in the answers:

a) Note that each of the eight values in part 1 is the sum of two values in part 2.

b) Because the 100 potatoes and 242 ears of maize are distributed according to the same scheme, the numbers of each per family are the same proportions of their respective wholes. The distribution scheme excludes the possibility of chopping up potatoes or maize. As a result, all the values are integers. The proportions are, therefore, "the same" to within that limitation. To demonstrate these proportions, our answer for part 1 is rewritten below with the number of potatoes per family divided by 100 and the number of ears of maize per family divided by 242.

Fig. 5.23

	Fam. 1	Fam. 2	Fam. 3	Fam. 4
Potatoes	0.300	0.350	0.100	0.250
Maize	0.302	0.351	0.099	0.248

c) Because each person's share is basically 5 potatoes and 12 ears of maize, and the family groups and age subgroups are sets of individuals, most of the data in the tables are multiples of 5 or 12.

d) Because, within each family, the potatoes and maize are distributed the same way, the ratios of items for adults to items for children is the same in both tables in part 2. This consistency can also be seen in the ratios of number of ears of maize to number of potatoes for each subgroup.

3. Here is one quipu schematic.

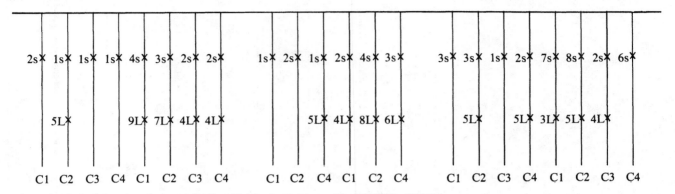

Colors C1, C2, C3, C4 are associated with families 1, 2, 3, 4 respectively. Cord group 1 is associated with items distributed to adults (potatoes followed by maize); cord group 2 is associated with items distributed to children; cord group 3 is associated with items distributed to whole families. (Check your quipu design to be sure that it conveys that there are no children in family 3. Everyone was given food—none received a share of zero).

Another observation: the data in the tables or on the quipu can be described as: a chart of elements p_{ij} ($i = 1, 2$; $j = 1, 2, 3, 4$) where each element is the sum of two elements in associated subcharts; subcharts of elements p_{ijk} ($i = 1, 2$; $j = 1, 2, 3, 4$; $k = 1, 2$); i is associated with food type, j with family, and k with age group.

Exercise 5.3

The first seventy-seven cords (quipu example 5.9) are separated into eleven groups. Each group consists of six pendant cords tied together with a top cord. From the color coding and the way the top cords are attached, we thought that relationships between values might become more apparent if some cord groups were read left to right and others right to left. The values are written in figure 5.25 in accordance with our thought. Notice that there is a pattern to the reversals so that the eleven groups form three sets: $(\leftarrow \rightarrow \rightarrow \leftarrow)$; $(\leftarrow \rightarrow \rightarrow \leftarrow)$; $(\rightarrow \leftarrow \rightarrow)$. Find any relationships you can within the groups. Find any relationships between groups. Find relationships between groups of groups. (If you wish to carry the problem further, rewrite the table so that all groups are read left to right and then look for other relationships.)

Fig. 5.24

Fig. 5.25. Quipu example 5.9.

							Top Cord
Group 1	15	24	42	12	37	16	146
Group 2	33	0	2	13	0	2	50
Group 3	33	34	161	37	44	18	327
Group 4	2	9	9	2	23	5	50
Group 5	97	44	57	214	46	55	513
Group 6	64	44	55	201	46	53	463
Group 7	17	0	71	21	14	73	196
Group 8	25	80	54	219	55	56	489
Group 9	99	54	67	232	64	71	587
Group 10	23	83	40	44	242	60	492
Group 11	100	85	110	433	90	104	922

Some Answers to Exercise 5.3

1. In every group, the value on the top cord is the sum of the values on the pendants.

2. Position by position, the values in group 2 plus the values in group 6 sum to the values in group 5.

3. For each of the first 5 positions, the sum of the values in the first 4 groups adds to a value in a later group in the next position.

That is, position 1 values in groups 1, 2, 3, 4 (15, 33, 33, 2) add to 83 which is in position 2 in group 10. Similarly,

a) position 2 values sum to position 3 in group 9;
b) position 3 values sum to position 4 in group 5;
c) position 4 values sum to position 5 in group 9;
d) position 5 values sum to position 6 in group 11.

4. $p_{54} + p_{84} = p_{64} + p_{94} = p_{11,4}$
5. $p_{52} = p_{62}$; $p_{55} = p_{65}$ (symmetric)
6. $p_{53} + p_{63} + p_{73} + p_{83} = p_{56} + p_{66} + p_{76} + p_{86}$
7. $p_{55} + p_{65} + p_{75} + p_{85} = p_{33}$

Chapter **6** Hierarchy and Pattern

1 We continue to explore logical structures by considering another prevalent quipu format. The basic construction features are still, of course, cord placement, color coding, and number representation. In the previous chapter, little was said about *subsidiary cords;* now, however, the fact that quipu cords can occur on different levels becomes important. In building a description of cross categorization, we used rectangular data tables, analogies to sets of tables and charts, and the more general concept of subscript notation. The structure of the format we now introduce is characterized by hierarchical categorization. Its description is easier. Tree diagrams, commonly used in describing branching processes or organization structures, provide the mode of visualization.

2 To record the monthly salaries of twenty people working for a company that sells shoes, the name of each person could be listed and an amount placed next to each name. However, more information is to be included in the record about the relationship of these people to the company and to each other. Suppose that the twenty people are the president of the company, four store managers, and fifteen salespersons. The salespersons are distributed so that three work at one store, three at another, and five and four at the third and fourth stores respectively. The relationships become visually explicit when the salary data is placed on a tree diagram reflecting this organization (fig. 6.1).

1 A belief in the importance of this type of categorization is expressed by L. L. Whyte as quoted in Thomas L. Saaty, *Hierarchies, Reciprocal Matrices and Ratio Scales,* Modules in Applied Mathematics (Washington, D.C.: Mathematics Association of America, 1976), p. 25: "The immense scope of hierarchical classification is clear. It is the most powerful method of classification used by the human brain-mind in ordering experience, observations, entities and information."

Fig. 6.1

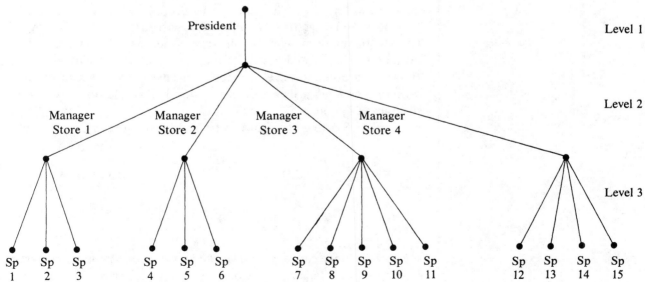

Notice that three levels are distinguished on the diagram: president, store managers, and salespeople. The total number of distinctions is the sum of the number of branches on each level $(1 + 4 + 15)$. Most important, starting at any branch on a lower level, there is one and only one way of tracing a path up to the root of the tree. (Incidentally, a tree diagram is actually an upside-down tree. The root is a single point at the top, and the tree spreads out its branches as you draw downward.)

If the data consisted of monthly salaries for a period of three years, there would be thirty-six such data trees, twelve for each of the three years. The thirty-six data trees could be easily arranged on a quipu. The quipu could have three groups of twelve pendants each. Each pendant would be the single branch on the first level of the tree. (The root of the tree is where the pendant is attached to the main cord.) Suspended from each pendant would be four subsidiary cords (the four branches on the second level of the tree), and these cords would have suspended from them three, three, five, and four subsidiaries respectively (the branches on the third level). The quipu would be a chart of elements p_{ij} ($i = 1, 2, \ldots, 12; j = 1, 2, 3$), where each element in the chart is a data tree.

To further examine tree diagrams, consider a different situation. A travel agent wishes to record, for presentation to his customers, the costs of travel from town A to town B to town C to town D. There are different prices for traveling first class or for traveling second class. But, in either case, three types of transportation are available: train, bus, or car. A choice of transportation has to be made from each town to the next. However, although a car can be dropped off anywhere, one can only be picked up at town A. Also, no buses run from town C to town D. Clearly, there are many different ways of putting these modes of transportation together and so, there are many different costs possible for the trip. In order to have the record of cost identified with the transportation options, a data tree can be used. First, let us draw the tree diagram of options.

Level 1: Starting at town A, a train, bus, or car can be selected.

Fig. 6.2

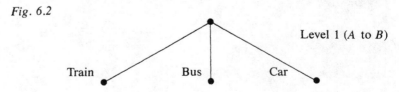

Level 1 (A to B)

Level 2: Since cars cannot be picked up at town B, only if one arrived by car, can a car be among the options. So, if one arrived by car, a train, bus, or car can be used. Otherwise, either a train or bus must be used.

Fig. 6.3

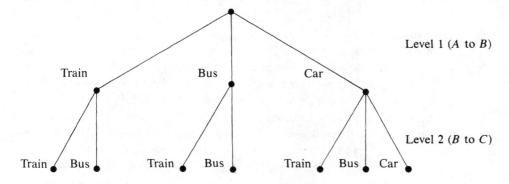

Level 1 (*A* to *B*)

Level 2 (*B* to *C*)

Level 3: From town *C*, there are no buses to town *D*. Also, the choice of a car is only open to those who arrived by car.

Fig. 6.4

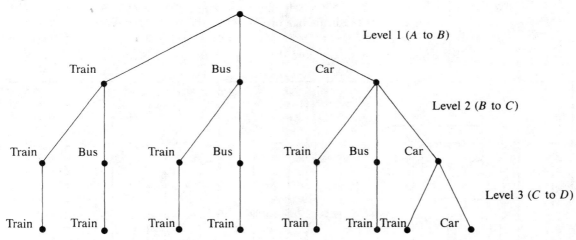

Level 1 (*A* to *B*)

Level 2 (*B* to *C*)

Level 3 (*C* to *D*)

This type of situation is called a branching process. The tree contains all the possible transportation combinations that are available to the customers. Each customer will select only one combination: for example, train followed by bus followed by train. It is a *process* because the overall selection is made up of consecutive choices; it is a *branching* process because a choice at one step leads to several possibilities when combined with the choices available at the next step.

The tree diagram has a feature that is quite different from the one reflecting organizational hierarchy. The same three choices, or a subset of them, are possible from each branch in moving from level to level. Thus, while individual labels are needed for every branch on the organizational tree, here only three labels are used repeatedly. Also, in an organizational tree, the number of branches on any level is completely arbitrary. In a tree diagraming a process with clearly stipulated choices at each level, the number of branches can be calculated. In the transportation tree, for example, if all three vehicles were available at each juncture, the three branches on level 1 would expand to 3×3 branches on level 2, and then to $(3 \times 3) \times 3$ branches on level 3.

As with the organizational example, the basic information unit is the tree diagram with numbers on it. In the organizational example, the numbers were salaries; here, the numbers are costs. Two data trees, both reflecting the same transportation options, but with different costs for the different classes of travel, constitute all the data that the travel agent wishes to record. A quipu containing the two data trees, one with the cost of first class travel and the other for second class travel, could have the layout shown in figure 6.5. Each group has three pendants, one for each of the branches on level 1 of the tree. Their colors are C1, C2, C3 associated with train, bus, and car respectively. Suspended from them are two, two, and three subsidiaries which correspond directly to

Fig. 6.5

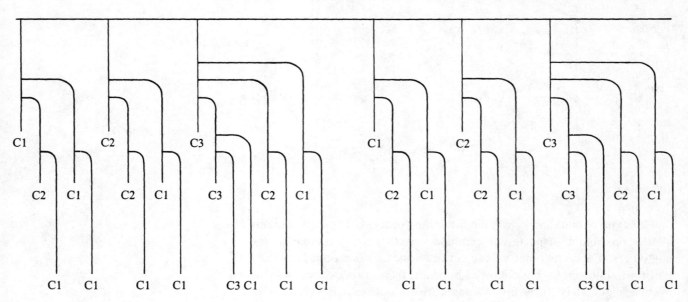

the level 2 branches on the tree. Their colors are also C1, C2, or C3 for train, bus, or car. Suspended from these subsidiaries are other subsidiaries. They correspond to the level 3 branches of the tree.

The overall description of this quipu would be a chart of elements p_{ij} ($i = 1, 2; j = 1, 2, 3$) where $i = 1$ and 2 correspond to the trees for first and second class travel respectively. Each tree has three levels with three branches on the first level; two or three branches from each of them form the second level; and one or two branches from each on level 2 form the third level. The branches on every level are selected from the same three colors.

3 Tree diagrams are particularly well suited to graphically describing organizations that emphasize lines of authority or processes in which choices are made successively. However, their use is not limited to these. They apply to any situation which can be described with consecutive levels and has one or more categories on each level, with the stipulation that there is a connection, and only one connection, between each category and the level above it. The description can be restated in terms of quipu components. There are consecutive levels: pendants, subsidiaries, subsidiaries of subsidiaries, etc. There can be one or more subsidiaries on any level. Each subsidiary is suspended from one cord and only one cord on the level above it, therefore satisfying the last stipulation. In hierarchical categorization, whether described by a tree diagram or a pendant and subsidiary array, the total number of distinctions being made is the total number of branches. But the logical complexity increases with both the number of levels and the number of branches on each level.

4 To elaborate more on the role of the relative placement of subsidiaries and color coding in hierarchical categorization, another example is considered. The data to be recorded is the number of hides distributed by ten hunters to members of their families at the end of a hunting season. The people receiving the hides in each hunter's family are categorized by generation, sex, and kinship, either direct or through marriage. The tree diagram for recording the data could look like figure 6.6. Generations have a consecutive order and so form the levels of the tree. The branches for children emanate from the parent directly related to the hunter so that there is only one connection per branch to the level above. Within the levels, from the second level downward, there is a branch for each of four categories: daughters-in-law, sons, daughters, sons-in-law.

A quipu for the distribution by the ten hunters would correspond to ten such data trees. It could have ten pendants, each with the subsidiary arrangement shown in the schematic in figure 6.7. The color C1 is associated with a hunter and C2, C3, C4, C5 with daughters-in-law, sons, daughters, and sons-in-law respectively. Therefore, a C5 subsidiary suspended from a C3 subsidiary suspended from a pendant cord corresponds to the sons-in-law of the sons of a hunter. If there were another level of subsidiaries, a C2 subsidiary suspended from a C3 subsidiary suspended from a C4 subsidiary suspended from a pendant cord would correspond to the daughters-in-law of the sons of the daughters of a hunter.

Notice that the subsidiaries from any branch, within each level, appear in the same color order: C2, C3, C4, C5. If all four subsidiaries were always present, these relative positions would be sufficient to identify the subsidiaries without color coding. However, there is no reason to expect that evey hunter will have children and grandchildren of all four categories. Some hunters may have only daughters; others may have no married sons. The categorization on the tree diagram (fig. 6.6)

Fig. 6.6

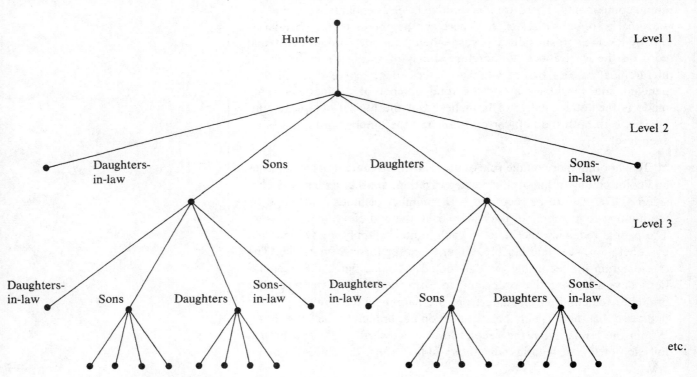

and on the corresponding quipu schematic (fig. 6.7) is of the most general form. The data tree for each hunter is a particular case of this general form. The color coding reinforces relative subsidiary placement so that there will be no ambiguity when there are only some of the subsidiaries present.

Although each data tree need not have all the subsidiaries of the general form present, there are some constraints imposed by the relationships within the data. If, for example, there is only one subsidiary present on level 2, the subsidiary must be color C3 or C4, assuming neither death nor divorce. That is, a hunter can have only sons or only daughters, but he cannot have only daughters-in-law or only sons-in-law. Similarly, if there are only two subsidiaries on level 2, they must be either (C3, C4), (C2, C3), or (C4, C5) and cannot be (C3, C5), (C2, C4), or (C2, C5). And, if there is only the pair (C3, C4) on level 2, no level 3 subsidiaries are expected.

The use of color coding to reinforce relative cord placement for hierarchical categorization is basically the same idea as was seen on quipus evidencing cross categorization. When examing actual quipus, if the

Fig. 6.7. One pendant cord with subsidiaries.

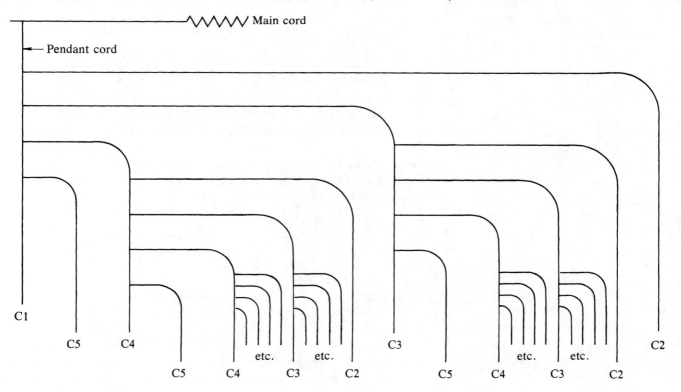

subsidiary arrays are all subsets of one data tree, the structure of the general tree can be identified. Further, noting which branches can be present or absent also gives information about the relationships within the data tree.

5 Quipu example 6.1 is AS62.
Quipu example 6.2 is AS94.
Quipu example 6.3 is AS52.
Quipu example 6.4 is AS69.
Quipu example 6.5 is AS181.
Quipu example 6.6 is AS70.

5 Subsidiary cords are present on about 65 percent of all quipus. However, not all quipus that have subsidiaries evidence hierarchical categorization. Some subsidiaries are used simply to associate more than one value with a cord position. Others are to record one value but in distinct parts. We regard a quipu as an expression of hierarchical categorization only when the subsidiary arrangement is consistent and repetitious. This occurs on about 20 percent of the quipus. The structures identified thus far (cross categorization and hierarchical categorization) are not mutually exclusive. They can be used separately or together, and when used together they build into one larger logical structure.

Several actual quipus are described in detail. To emphasize their logical structures, they are depicted as tree diagrams. The first two of the examples display hierarchical categorization alone; the next two combine hierarchical and cross categorization. Then a quipu example of the previous chapter is continued; it includes hierarchical categorization, cross categorization, and summation.

It is difficult to imagine keeping in mind all the categories and relationships between categories. The last two examples are included because they are impressive in their complexity. But it is even more difficult to imagine any other device so aptly suited to the recording and visual display of the complex logical relationships between recorded numbers.

Quipu Example 6.1

The quipu consists of eight data trees. Each tree has all branches. A typical tree is shown in figure 6.8.

Fig. 6.8. Quipu example 6.1.

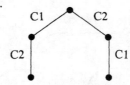

Quipu Example 6.2

The quipu consists of one data tree. The tree is shown in figure 6.9. All six level 1 branches are C1. From each, six level 2 branches emanate. With one exception they are C1, C1, C1, C2, C1, C1. Again with one exception, all level 3 branches are C3.

CODE OF THE QUIPU

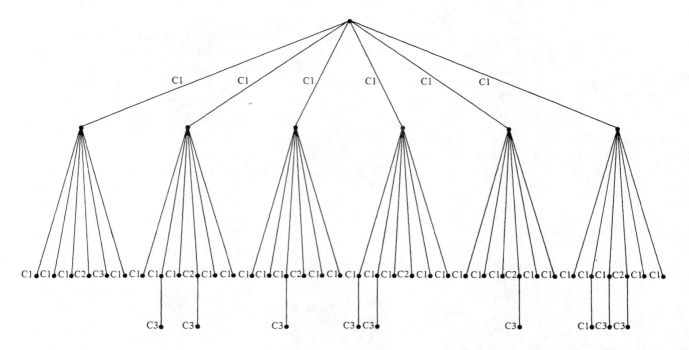

Fig. 6.9. Quipu example 6.2.

Quipu Example 6.3

This example combines hierarchical and cross categorization.

    1. Chart with elements p_{ij} ($i = 1, 2, 3, 4$; $j = 1, 2, 3, 4, 5$).

    2. Each element p_{ij} ($i = 1, 2$; $j = 1, 2, 3, 4, 5$) is associated with a data tree. The general form of the tree is shown in figure 6.10A. The particular trees have either zero, one, or two branches on level 2. These are never C3 alone or (C1, C3).

    3. Each element p_{ij} ($i = 3, 4$; $j = 1, 2, 3, 4, 5$) is associated with a data tree. The general form of the tree is shown in figure 6.10B. All the particular trees have (C1, C2) or C2 alone on the second level.

Fig. 6.10. Quipu example 6.3.

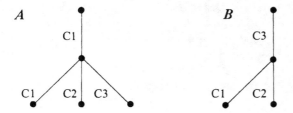

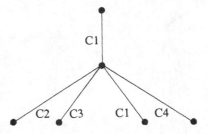

Fig. 6.11. Quipu example 6.4.

C1

C2 C3 C1 C4

Quipu Example 6.4

This quipu is remarkable in its size and excellent state of preservation (see plate 6.1). It is, without doubt, currently the largest quipu in the world. Careful planning and careful workmanship are evident on it. With this quipu were found three other quipus and prepared pendant blanks. Plates 2.1, 2.2, and 2.3 are from the set as is quipu example 6.6 of this chapter. The use of color to identify pendant groups and subgroups can be seen on the quipu.

1. Chart of elements p_{ijk} ($i = 1, \ldots, 9$; $j = 1, \ldots, 27$; $k = 1, \ldots, 7$).

2. Each p_{i5k} ($i = 1, \ldots, 9$; $k = 1, \ldots, 7$) is associated with a data tree whose general form is shown in figure 6.11. The particular trees have zero, one, two, or four branches on level 2. When there is only one branch on level 2, it is always C2; when there are two branches they are (C2, C3) or (C1, C4).

Plate 6.1. Quipu example 6.4. *(In the collection of P. Dauelsberg, Arica, Chile.)*

CODE OF THE QUIPU

Quipu Example 5.2—*Continued*

1. Chart elements p_{ij} ($i = 1, 2; j = 1, \ldots , 18$) and subchart elements p_{ijk} ($j = 1, \ldots , 18; k = 1, 2, 3$) are each the single branch on the first level of a data tree. The second level has, in general, eleven possible branches (C1, C2, C3, C4, C5, C3, C5, C6, C7, C8, C9). No particular tree has all the branches (and some have a few more C3 or C9).

2. On level 2, branches C1, C7, and C8 are only present for $j = 1$ or 10. Branch C4 is only present for $j = 14$ or 15.

3. The level 2 branches of elements p_{1j} ($j = 1, \ldots , 18$) have numerical values which are the sums of the values on the corresponding branches of p_{2j}, p_{1j1}, p_{1j2}, and p_{1j3}. Since p_{1j} does not have level 2 branches, C1, C2, C4, their sum is not included.

4. This is the quipu shown in plate 2.6. On the plate can be seen the six groups of eighteen pendants. The first group is the categorical sums; groups 2 and 3 are the chart; and groups 4, 5, and 6 are the subchart. The subsidiaries are also visible. In fact, one can observe some of the color differences and that there are more subsidiaries on the first and tenth pendants than on the others.

Quipu Example 6.5

1. Ten consecutive pendants are associated with five data trees. The general form of the tree is shown in figure 6.12*A*. The circled branches are on only one particular tree. On that same tree, C6 on level 4 is absent.

2. Another ten consecutive pendants are associated with five other data trees. Their general tree is shown in figure 6.12*B*. C11 on level 2 is absent on two particular trees.

Fig. 6.12. Quipu example 6.5.

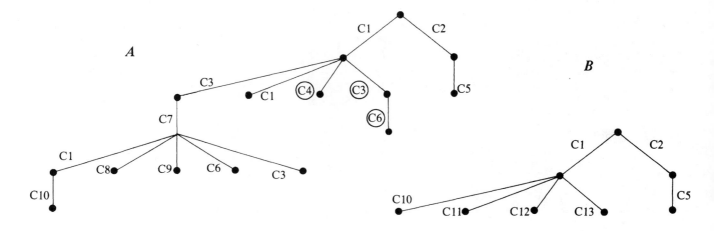

Quipu Example 6.6

1. Chart of elements p_{ij} $(i = 1, \ldots , 8; j = 1, \ldots , 10)$.

2. Each element p_{ij} $(i = 1, \ldots , 7; j = 9, 10)$ is associated with a data tree. The general form of the tree is shown in figure 6.13. Note that the branches on level 3 are the same as those on level 2 but with one fewer C1. Similarly, the branches on level 4 are the same as those on level 3 with still one fewer C1. No particular tree has all branches; the maximum number of branches on a tree is nine.

3. Each element p_{ij} $(i = 1, \ldots , 7; j = 7, 8)$ is associated with a data tree. The general form of the tree includes all the branches of figure 6.13 as well as two more branches from each juncture. Hence,

level 2 branches are C1, C1, C2, C4, C3, C5;

level 3 branches are C1, C2, C4, C3, C5; and

level 4 branches are C2, C4, C3, C5.

No particular tree has all branches; the maximum number of branches on a tree is fourteen.

4. Additional elements conform to the latter tree with some variation. For example, p_{65} also has a fifth level with a single C4 branch. In total, it has sixteen branches.

5. Although not with the same consistency of structure, other elements have as many as ten branches on the second level and some have as many as six levels. In all, twenty-five of the trees on this quipu have ten or more branches. Of these, five trees have fifteen to twenty branches and two have twenty-one to twenty-two branches.

Fig. 6.13. Quipu example 6.6.

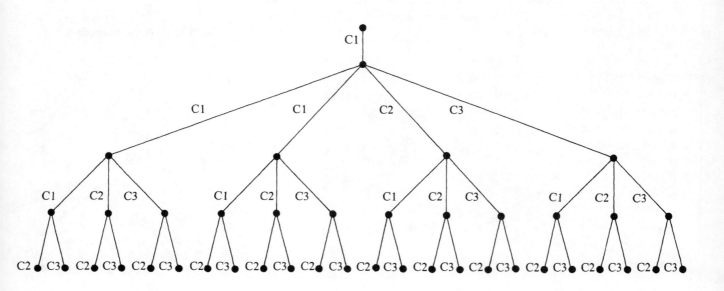

CODE OF THE QUIPU

6 Before continuing with other structures, an important gap in hierarchical categorization must be filled. The tree diagram model evades a problem that is encountered in reality. In the organizational example, there were three levels: president, store managers, and salespeople. The branches for the salespeople at each store emanated from a branch for a manager on the level above. But, what if a store is without a manager? Similarly, in the example of the distribution of hides by a hunter, we arbitrarily ruled out death and divorce of his children thus insuring that his grandchildren on level 3 were connected to a branch on level 2. The problem, in both cases, is how to keep level 3 on level 3 when there is no level 2 above it. The solution most frequently encountered on tree diagrams are dotted lines or other special notations explaining the deviation from the model.

At least one quipumaker solved the problem using differential cord placement. On a quipu with data trees with three levels, the subsidiaries forming the various branches on the second level were suspended from the pendant cords between 3.0 and 8.0 cm. down from the main cord. However, on one pendant, a single subsidiary was suspended at 15.0 cm. Its color and placement clearly conveyed that it was a third level branch, although there were no second level branches present. Thus, we have another instance where space on a cord has significance. This gives the quipu the flexibility needed to express a slightly modified form of hierarchical categorization.

7 The formats discussed thus far are of importance because they subsume many quipu specimens. However, also needed in the exploration of structures, is consideration of individual quipus that are different from these. As we move to the consideration of individual quipus, the manner of their examination also changes. We do not attempt to relate the individual quipus to other quipus; we are simply interested in the patterns the particular quipus display. The two approaches are somewhat analogous to two different modes of literary analysis. One can classify written works as, say, plays or novels and discuss the commonalities of structure for each. But the examination of an individual poem for how it says what it says, extends the understanding of what can be done with words when they are put together. When an individual poem is studied by itself, it is not because of an insufficiently large sample of poems. In each, something can be found that need be found in no other to be of importance.

In the same sense, we now consider individual quipus with different structures. Here we look at quipus that use numbers as labels rather

6 The quipu referred to is AS31. Another that possibly shows use of the same technique is AS150.

than as numerical data. Later, in another chapter, we return to numerical data but continue with the examination of individual quipus. There, too, one instance of a clear statement is sufficient to place the statement within the body of ideas used by the Incas. In both cases, the usages of fundamental construction features (cord placement, color coding, number representation) that have already been established are important prerequisites. They are used as guides and constraints in the examination of individual quipus.

8 The scope of recording is greatly enlarged when knots are interpreted as number labels mixed with magnitudes rather than solely as magnitudes. There are, in our contemporary Western environment, many schemes routinely used and familiar to their users, that combine labels and magnitudes. In written and printed records, however, the labels are more commonly letters and other typewriter characters. A few examples to illustrate these are:

a) CH_3Cl, C_2H_5Cl, $C_{10}H_{17}Cl$, $ClCOOC_2H_5$ $\}$ chemical compounds

b) Em, B7, Em, D7, G, E7, Am, B7, Em,
   E7, Am, E7, Am, B7, Em $\left.\right\}$ guitar chords

c) R1: P4, *K1, P7*, E K1, P4.
   R2: K4, *P1, K7*, E P1, K4. $\left.\right\}$ knitting instructions
   R3: P3, *K3, P5*, E K3, P3.

Two of the individual quipus are described as rhythmic. Others have patterns that could be termed symmetric or variations on a theme. These terms for elements of structure are often applied to what scientists find in nature or what creative artists attempt to achieve. The quipus that show these patterns are, of course, recordings. We suggest that their structures match the phenomena recorded, whether they be tunes, plans for textiles, or phases of the moon. The numerical data discussed earlier were ordered by the formatization imposed by the quipumaker. As a designer of formats, the quipumaker's role was the active and imaginative one of achieving a recording that indeed reflected the essence of the phenomenon he recorded. The examples following are our rendering of quipus with the aim of conveying what we perceive as important in them.

9 Quipu Example 6.7

This quipu is a simple rhythmic arrangement of number labels and magnitudes. Knot placement and color coding make the distinction between labels and magnitudes evident. Cord grouping, relative placement of groups, and relative placement of pendants within the groups convey the rhythm. Webster's dictionary includes definitions of rhythm in music

8 (a) The chemical compounds are methyl chloride, ethyl chloride, bornyl chloride, and ethyl chloroformate.

(b) The guitar chords are for "Zog Nit Keymol" as arranged by Robert DeCormier in *Lift Every Voice!* (New York: People's Artists, 1953), p. 52.

(c) The knitting instructions are adapted from *Bernat Handicrafter*, Book no. 67, ed. C. Goldsmith (Jamaica Plain, Mass.: Emile Bernat and Sons, 1958), p. 41.

9 (a) The definitions are from *Webster's New World Dictionary of the American Language,* College ed. (Cleveland, Ohio: William Collins and World Publishing Co., 1966), p. 1251.

(b) Quipu example 6.7 is AS145. Quipu example 6.8 is AS101.

and rhythm in prosody. Compare them before examining the quipu configuration in figure 6.14.

In music: regular (or occasionally, somewhat irregular) recurrence of grouped strong and weak beats, or heavily and lightly accented tones, in alternation; arrangement of successive tones, usually in measures, according to their relative accentuation and duration.

In prosody: basically regular recurrence of grouped, stressed and unstressed, long and short, or high-pitched and low-pitched syllables in alternation; arrangement of successive syllables, as in metrical units (feet) or cadences, according to their relative stress, quantity, and pitch.

Fig. 6.14. Quipu example 6.7.

L1, L1, L1, L1, L1, L1, L1
N1, N1, N1, N1, N1, N1, N1
L1, L1, L1, L1, L1, L1, L1
N1, N1, N1, N1, N1, N1, N1
L2, N2, L2, N2, L2, N2, L2, N2, L2, N2, L2, N2, L2, N2
L2, N2, L2, N2, L2, N2, L2, N2, L2, N2, L2, N2, L2, N2
L1, N2, L2, N2, L2, N2, L2, N2, L2, N2, L2, N2, L1, N2, L1, N2
L1, N2, L1, N2, L2, N2, L2, N2, L2, N2, L2, N2, L1, N2

Note:
N1 = cords with standard arrangement of single knots and long knots; interpreted as magnitudes; color C1.
N2 = cords with standard arrangement of single knots and long knots; interpreted as magnitudes; color C2.
L1 = cords with only long knots in all knot clusters; interpreted as labels; color C3.
L2 = cords with only long knots in all knot clusters; interpreted as labels; color C4.

The first four groups are separated from each other by space along the main cord.
The second four groups are separated from each other by space along the main cord.
The first four groups are separated from the second four groups by a larger space.

Quipu Example 6.8
This example is, actually, two quipus that were tied together. Individually they exhibit rhythmic patterns. Moreover, the pattern of one is a variation of the other. And, the ways in which the pattern is exhibited are the same for both. We will call the quipus in the set quipu I and quipu II.

(*c*) The knot clusters on each cord cannot necessarily be interpreted as base 10 positional numbers. Therefore, where a cord has more than one knot cluster, the number of knots in each has been listed with a hyphen between clusters. For example, 1–6 is a cluster of one single knot followed by a cluster of six single knots. (Notice that values 2, 3, . . . , 9 are found where there is only one cluster on a cord. Where there are two clusters, the first is a 1 and the second is 2, 4, 6, or 8.)

Fig. N.12. Quipu example 6.8.

Quipu I

Pendants in group 1	1–6, 1–8, 1–6
Subsidiaries in group 1	8, 1–4, 8
Pendants in group 2	3, 4, 3
Subsidiaries in group 2:	
higher	3, 4, 3
lower	1–6, 1–8, 1–6
Pendants in group 3	3, 4, 3
Subsidiaries in group 3	1–6, 1–8, 1–6

Quipu II

Pendants in group 1	3, 3, 7, 3
Subsidiaries in group 1	2, 2, 5, 2
Pendants in group 2	7, 7, 9, 7
Subsidiaries in group 2	3, 3, 4, 3
Pendants in group 3	3, 3, 7, 3
Pendants in group 4	6, 6, 1–2, 6
Subsidiaries in group 4	4, 4, 6, 4

(*d*) Quipu Example 6.9 is AS15. The other two quipus with tails are AS27 and AS82.

Quipu I has a persistent *aba* pattern. The pattern is in the overall cord arrangement and then within that in the juxtaposition of the cord knot values. Specifically,

1. The quipu consists of three groups of cords: the first group has three pendants with one subsidiary each; the next has three pendants with two subsidiaries each; and the third has the same arrangement as the first.

2. Within each group, the knot values on the pendants are the same for the first and third pendants and different for the middle pendant.

3. Within each group, the knot values on the subsidiaries are the same for those suspended from the first and third pendants and different for the middle pendant.

In the same ways, quipu II has a persistent *aaba* pattern.

1. The quipu has four groups of cords: the first group has four pendants with one subsidiary each, as do the second and fourth groups. The third group is different as it has no subsidiaries.

2. Within each group, the knot values on the pendants are the same for the first, second, and fourth pendants and different for the third pendant.

3. Within each group, the knot values on the subsidiaries are the same for those suspended from the first, second, and fourth pendants and different for the third pendant.

We were struck by the fact that the colors played no role in the patterning on each quipu but then found that the colors formed another linkage between quipus I and II. By color, the groups of cords are isomorphic: that is, the groups of cords of quipu I can be matched, one for one, with the groups of quipu II. The correspondences show how the *aba* cord arrangement is transformed into the *aaba* arrangement (see fig. 6.15).

Fig. 6.15. Quipu example 6.8.

	Quipu I		Quipu II	
A	Pendants of group 1	←————→	Pendants of group 1	A
	Subsidiaries of group 1	←————→	Subsidiaries of group 1	
B	Pendants of group 2	←————→	Pendants of group 2	A
	Higher subsidiaries of group 2	←————→	Subsidiaries of group 2	
	Lower subsidiaries of group 2	←————→	Pendants of group 3	B
A	Pendants of group 3	←————→	Pendants of group 4	A
	Subsidiaries of group 3	←————→	Subsidiaries of group 4	

The belief that this is not simply a numerical record is corroborated by the nonstandard use of knots. Only single knots are used in all knot clusters. Therefore, all or some may be number labels. The values of the knot clusters are included in the chapter notes (p. 123) in case the reader wishes to continue the search for patterns on these quipus. It would be particularly interesting if rules of correspondence could be found that would transform the knot values on quipu I into those on quipu II.

Quipu Example 6.9

This is a most intriguing quipu. A long cord suspended from a loop attached to the main cord can be seen on plate 6.2. This "tail" is clearly neither a pendant cord nor a dangle end cord. Small pieces of cord are

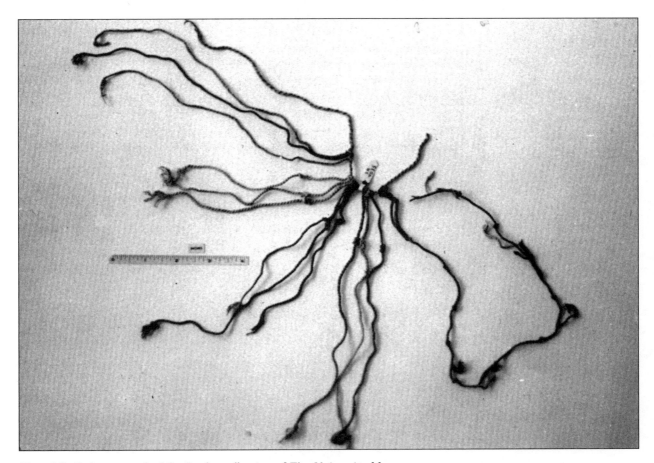

Plate 6.2. Quipu example 6.9. (*In the collection of The University Museum, University of Pennsylvania, Philadelphia.*)

attached to it. Because of the way they are attached, and because they are much too short to have knots on them, these are not subsidiaries but are "flags" of some sort. This construction component, a tail with single knots interspersed with colored flags, was found on only two other quipus. Another striking aspect of the quipu is that each pendant has either one figure eight knot or no knots at all. Rather than representing the numbers 1 and 0, these could represent any dichotomy such as yes/no, present/absent, or in/out. The quipu has, in all, pendants colored C1, C2, C3, C4 or C5, each with a knot (K) or no knot (0), in the following order: C1(K), C2(K), C1(K), C1(K), C1(K), C1(K), C2(K), C2(K), C2(K), C3(0), C4(0), C5(0); and a tail with flags of color C1 or C2 separating the single knots (s) into groups of one, two, or three in the following order: 3s, C1, 1s, C2, C1, 2s, C1, 1s, C2, C2, C1, 2s, C2, C2.

Comparison of the body of the quipu with its tail shows a basic theme restated in different ways. There are nine knotted pendants and there are nine flags. They can be placed in a one-to-one correspondence with each other because they are in exactly the same color order. There are also nine single knots on the tail. Just as the flags separate the knots into groups, the knots separate the flags into groups. The numbers of knots in the knot groups are 3, 1, 2, 1, 2. The sizes of the flag groups are 1, 2, 1, 3, 2. The sizes of the groups are the same, and the orders in which they occur correspond to each other according to the rule:

size of $(5-i)^{th}$ flag group = size of ith knot group
$i = 1, 2, 3, 4, 5$; arithmetic is mod 5.

To visualize the rule above, view the tail as being tied in a circle and list the knot group sizes clockwise:

Fig. 6.16

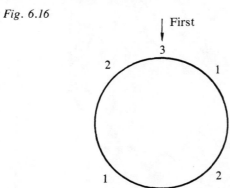

rotate the circle two positions clockwise:

Fig. 6.17

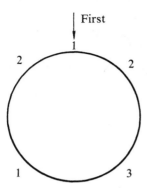

and now read off the sizes counterclockwise: 1, 2, 1, 3, 2.

Thus, there are essentially nine elements. They are associated with the color order C1, C2, C1, C1, C1, C1, C2, C2, C2 as expressed on the flags and reiterated on the knotted pendants. The nine elements are divided into groupings of 1, 2, 1, 3, 2 as expressed on the flags and reiterated by the knots. The relative order of the groupings remain the same whether they continue around or are listed in either direction. It is, therefore, the relative, rather than absolute order, that is being emphasized by the reiteration.

10 The information on any quipu is in both the data and the organization of the data. Planning, design, and construction, which embedded information into a quipu, had to be complemented by retrieval of the information. Recognizing a quipu as being of a prevalent format provides the conceptual framework for retrieving information from it. Interpreting and understanding quipus whose structures are more individualized also has to draw on a set of shared ideas. Prevalent formats and individualized quipus that are highly ordered compositions result from the same underlying system of concepts and rules for the expression of the concept.

EXERCISES

Exercise 6.1

1. Draw a tree diagram to represent the following game. A box contains three cards which are colored red, yellow, and green. The player draws a card and receives $1.00 if red was drawn, $2.00 if yellow was drawn, and nothing if green was drawn. The card is returned to the box and the player draws a card again. He receives $4.00 more if he draws the same color as he did the first time. He

Exercises (*a*) The afghan of exercise 6.2 is B–894 of *Creative Afghans*, Coats and Clark's Book no. 223 (New York, 1972). The assembly diagram is on page 7.

(*b*) The quipu in exercise 6.3 is AS98. The second pendant in groups 14 and 15 each have a subsidiary of color C6. For simplicity, this was omitted from the exercise.

receives nothing and must stop playing if he draws a different color. The card is again returned to the box. On the third and last draw, the player gets $10.00 more if he draws the same color and nothing for a different color.

2. Design a quipu to represent the same situation.

An Answer to Exercise 6.1

1. Tree diagram

Fig. 6.18

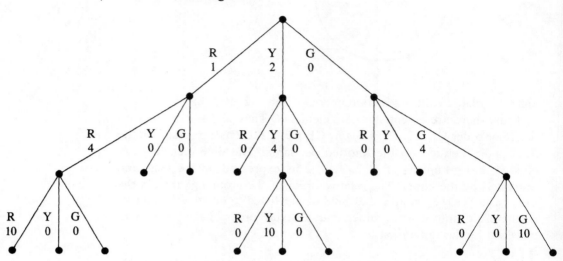

Fig. 6.19   2. Quipu schematic

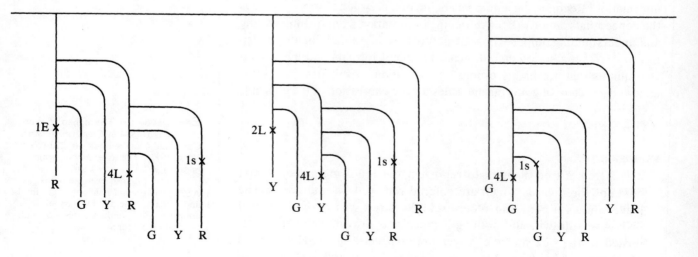

CODE OF THE QUIPU

Exercise 6.2

The making of an afghan includes knitting squares of four different solid colors and embroidering some of them with a leaf motif. The scheme for assembling the finished squares is diagramed in the instructions.

Fig. 6.20. D = embroidered motif.

1–D	3	2–D	4	3–D
2	4–D	3	1–D	2
3–D	1	4–D	2	3–D
1	2–D	3	4–D	1
2–D	4	1–D	2	1–D
1	3–D	2	4–D	3
2–D	4	3–D	1	2–D

1. Design a quipu for displaying the assembly scheme. Give the answer in the form of a schematic.

2. Describe any patterns you can find in the planned afghan.

An Answer to Exercise 6.2

1. The quipu could have four colors of pendants C1, C2, C3, C4 corresponding to the four colors of knitted squares. The dichotomy, embroidered/not embroidered, can be represented by a knot or no knot. Each group of pendant cords corresponds to one row on the diagram.

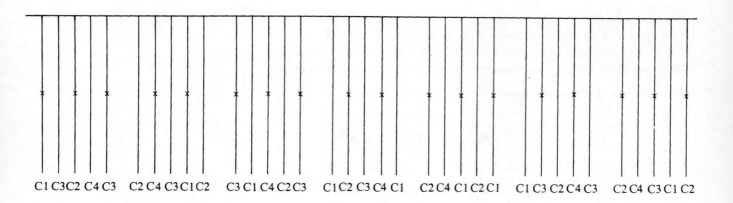

C1 C3 C2 C4 C3   C2 C4 C3 C1 C2   C3 C1 C4 C2 C3   C1 C2 C3 C4 C1   C2 C4 C1 C2 C1   C1 C3 C2 C4 C3   C2 C4 C3 C1 C2

Fig. 6.21

2. Embroidered motifs alternate within each row. However, in alternate rows they begin at the edge or one square over. Therefore, rows 1, 3, 5, 7 have DODOD and the alternative rows 2, 4, 6 have ODODO. Also, the first two rows and last two rows are the same colors: the color order in row 1 corresponds to the color order in row 6, and that of row 2 corresponds to row 7.

Exercise 6.3
A quipu currently housed in Berlin consists of fifteen groups of pendant cords. Groups 1, 4, 7, 10 and 13 have six pendants each. Groups 2, 3, 5, 6, 8, 9, 11, 12, 14, and 15 each have five pendants with two subsidiaries on the first. The colors of cords in groups 1, 4, and 7 are C1, C2, C2, C2, C2, C2 and those in groups 10 and 13 are all color C5. The colors of cords in groups 2, 3, 5, 6, 8, 9 are C2 (with subsidiaries of C1 and C2), C3, C2, C4, C2 and those in groups 11, 12, 14, and 15 are C5 (with subsidiaries of C5, C5), C3, C5, C4, C5.

1. Find a rule that transforms the colors of group 2 into those of group 11, and which also transforms the colors of group 1 into those of group 10.

2. Let X stand for the cord arrangement and color pattern of group 1 and X' stand for the same cord arrangement but transformed color pattern in group 10. Similarly, let Y stand for the arrangement and color pattern of group 2 and Y' stand for the same arrangement but transformed color pattern in group 11. Using only the symbols X, X', Y, Y', write a description of the fifteen cord groups on the quipu.

3. Using your own terminology, describe the overall pattern that results from 2.

130

Answer to Exercise 6.3

1. Every place that group 2 has colors C1 or C2, group 11 has color C5. Similarly, every place that group 1 has C1 or C2, group 10 has color C5. The rule of transformation is, therefore, replace all C1 and C2 by C5.

2. XYYXYYXYYX'Y'Y'X'Y'Y'

Chapter 7

Arithmetical Ideas and Recurrent Numbers

1 Numbers with their sums have already been seen on many quipus. But there are other numerical relationships on quipus that show other arithmetical ideas. In order to understand these ideas, we need to discuss more about numbers and their representations.

2 How positive integers are symbolized in our base 10 positional system and how integers are represented by knots on quipus has already been explained. But we write more than just positive integers. If, for example, one wants to represent one-half, it is written as the fraction 1/2 or as the decimal fraction 0.5. These symbols are in different notational systems yet they represent the same value. A fraction is a single entity but contains two distinct integers (for example, 2/3, 7/4, 1/324). The value of a fraction can be more or less than 1. A decimal fraction (0.372, 0.25, 0.414141. . .), on the other hand, is part of the base 10 positional notation. In base 10, a decimal point is used to separate integers from decimal fractions. Starting at the decimal point, as you move to the left, the value of each symbol in the collection is multiplied by one higher power of 10; starting at the decimal point, as you move to the right, the value of each symbol is *divided* by one higher power of 10. For example, 437.243 has the value

$$(4 \times 100) + (3 \times 10) + (7 \times 1) + 2/10 + 4/100 + 3/1000.$$

Every fraction can be translated into base 10 positional notation but not every decimal number can be translated into a fraction. Most people have internalized the translation from one notation to the other for a limited set of values, such as 0.5 = 1/2, 0.75 = 3/4, 0.333. . . = 1/3. (The means of translation from one notation to the other is elaborated in the notes that accompany the chapter. Which decimal numbers cannot be translated and what we do with them, is also included there.)

Since decimal numbers can represent any value that a fraction can, we could just use them but we persist in using both. This is probably because they stress different ideas and so are useful in different ways. Fractions give a direct statement of the result of dividing some number into several equal parts. For example, when 6 is divided into 8 equal parts, each resultant part is simply 6/8. Also, viewing a fraction as a ratio, one can find from it other numbers with the same ratio. For example,

$$6/8 = 2(3)/2(4) = 3/4 \quad \text{or} \quad 6/8 = 3(6)/3(8) = 18/24.$$

Decimal fractions, on the other hand, greatly simplify arithmetic. Each fractional value has the same general representation: the parts of

2 Every fraction can be translated into base positional notation. The translation procedure is long division. That is, to find the decimal representation of 3/4, divide 4 into 3:

$$
\begin{array}{r}
0.75 \\
4\overline{\smash{\big)}\ 3.0} \\
2\ 8 \\
\hline
20 \\
20 \\
\hline
\end{array}
$$

However, there are some values that can be represented in base 10 positional notation that cannot be translated into fractions. One can tell, just by looking at the representation, if it is translatable.

If the decimal fraction has a finite number of positions, it can be written as a fraction. Examples of decimal fractions with 1, 2, and 4 positions and their translations are:

$$0.5 = \frac{5}{10} = \frac{1(5)}{2(5)} = \frac{1}{2};$$

$$0.75 = \frac{7}{10} + \frac{5}{100} = \frac{70+5}{100} = \frac{75}{100} = \frac{3(25)}{4(25)} = \frac{3}{4};$$

$$0.6216 = \frac{6}{10} + \frac{2}{100} + \frac{1}{1,000} + \frac{6}{10,000} = \frac{6,000 + 200 + 10 + 6}{10,000} = \frac{6,216}{10,000} = \frac{8(777)}{8(1,250)} = \frac{777}{1,250}.$$

A decimal fraction that has an infinite number of positions but in which, from some position on, the same digits are repeated in a cycle, can also be written as a fraction. For example,

0.3333 . . . = 1/3;
0.2143143143 . . . = 2,141/9,990.

To translate an infinitely repeating decimal fraction into a fraction, the number is multiplied by powers of 10 such that when the multiples are subtracted from each other only an integer remains. Integer arithmetic is then used.

Example: $x = .2143143 \ldots$ 1. Let x stand for the value.

$10x = 2.143143 \ldots$ 2. Multiply by the power of 10 that places the first cycle at the decimal point.

$10,000x = 2,143.143143 \ldots$ 3. Multiply by the power of 10 that places a single cycle to the left of the decimal point.

$9,990x = 2,141$ 4. Subtract the result of step 2 from the result of step 3.

$x = \dfrac{2,141}{9,990}$ 5. Divide and reduce if possible.

Example: $x = .666 \ldots$ Step 1.
There is no need for Step 2.
$10x = 6.666 \ldots$ Step 3.
$9x = 6$ Step 4.
$x = 2/3$ Step 5.

This leaves decimal numbers whose fractional parts have an infinite number of positions but which do not cycle. There is no way of translating them into fractions. This is not because we do not know the translation rule; it is because no fraction can have an equivalent value. These numbers, incidentally, are inappropriately, but perhaps understandably, called irrational numbers. When such a number commonly occurs because it corresponds to a value that arises in a common situation, it is often represented by a special sign which is not in either of the two notational systems. The most familiar is π. The value of the ratio of the circumference of a circle to the diameter of that circle is an irrational number. There is no integer that when divided by another integer is equivalent to it, so it cannot be represented as a fraction. If one were to write it as a decimal, the representation is unending and, hence, most inconvenient. Any decimal representation that uses only a finite number of digits, such as 3.1416, is only an approximation to the value. Also, any fraction representation such as 22/7, is only an approximate value.

CODE OF THE QUIPU

numbers that are used are confined to powers of 10. Thus, of all the fractions that are equivalent to say, 6/8, one uses 75/100 or 0.75. With this systematized way of representing all fractional values, arithmetic can be done through their systematic manipulation. What is more, the manipulations are the same as for integers.

3 Clearly, the idea of dividing things into fractional parts and the idea of common ratios can exist with or without a scheme for the representation of fractions. And fractional parts can be combined and modified in systematic ways that do not involve manipulations of symbolic representations. On the simplest level, even a child who has no knowledge of how to write fractions can divide up a set of items. But complex mathematics, too, can be built without fractional representation. The Greeks of the fourth century B.C.E. and throughout what is often referred to as the golden age of Greek mathematics (first century C.E.) are a good illustration of this. They had no general way of representing fractional values yet they, nevertheless, devoted considerable attention to number theory and to ratios. Their approach, however, was theoretical and geometric rather than computational. The Greeks not only had no representation of fractional values, they had a definite belief that there should be none.

On quipus we find no evidence of the representation of fractional values. We do find evidence for division into parts and the use of common ratios. Just as in the case of summation, where numbers and their sums appeared together, we have no way of knowing how or where the calculations were done or even what calculations were done.

4 Several quipus show the arithmetic idea of even division into parts. The simplest example is a quipu where the total of all the values is 200. The quipu consists of two cord groups, each of which has a total value of 1/2 of 200 (100). Then, in the cord group made up of two pendants, each has 1/2 the value of the total for the group (50). The other group has six pendants. The values are the result of dividing 100 evenly into six parts.

Consider, for a moment, the results of dividing integers with the restriction that only integers can occur. If 12 is divided evenly into three parts, the result is 4, 4, and 4. If 13 is divided evenly into three parts, the result cannot be 4 1/3, 4 1/3, 4 1/3. It must be 4, 4, 5. This manner of division is frequently used by us in situations where we restrict ourselves to integral results even though fractions are available. People are viewed as units with integrity. We view many other things as funda-

3 (*a*) Florian Cajori, *A History of Mathematical Notations,* vol. 1 (1928; reprint ed., LaSalle, Ill.: Open Court Publishing Co., 1974), contains examples of expressions of fractions in words by the Greeks. For example, Eratosthenes' description of 11/83 of a unit arc of the earth's meridian is "amounts to eleven parts of which the meridian had eighty-three" (about 200 B.C.E.). Also, he notes that unit fractions were frequently used by Geminus (first century B.C.E.) and later by Diophantus (third century C.E.) and Eutocius and Proclus (fifth century C.E.), and that "the fraction 1/2 had a mark of its own, namely ι or ι' , but this designation was no more adopted generally among the Greeks than were the other notations of fractions." See his pages 26–27.

(*b*) To give some idea of how ratios were dealt with without fractions, we quote from Carl B. Boyer, *A History of Mathematics* (New York: John Wiley and Sons, 1968). Prior to Eudoxus "Apparently the Greeks had made use of the idea that four quantities are in proportion, $a:b = c:d$, if the two ratios $a:b$ and $c:d$ have the same mutual subtraction. That is, the smaller in each ratio can be laid off on the larger the same integral number of times, and the remainder in each case can be laid off on the smaller the same integral number of times, and the new remainder can be laid off on the former remainder the same integral number of times, and so on" (pp. 98–99). Then, from Euclid's Book V which deals with the general theory of proportions, he states Eudoxus' definition. "Magnitudes are said to be in the same ratio, the first to the second and the third to the fourth, when, if any equimultiples whatever be taken of the first and the third, and any equimultiples whatever of the second and fourth, the former equimultiples alike exceed, are alike equal to, or are alike less than, the latter equimultiples taken in corresponding order" (p. 99).

(*c*) The Greek view that there should be no fractional representation is clearly set forth in Plato's *The Republic* (Book VII, 525, as translated by Benjamin Jowett). ". . . steadily the masters of the art repel and ridicule anyone who attempts to divide absolute unity when he is calculating. . . ."

mental or indivisible units. For example, 9 pennies evenly divided into two parts must result in 4 pennies and 5 pennies.

Since 100 is not an exact multiple of 6, the result of dividing it evenly into six parts does not lead to equal values. It results in 16, 16, 17, 17, 17, 17. These are the values found on the quipu in the example.

Another quipu has cord groups each of which have eight pendant cords and a top cord. The values on the top cords are the sums of the values on the pendants in the associated groups. In several groups, where the sum is an exact multiple of 8 (32, 40, 104, 192), each pendant has a value that is 1/8 of the sum. Where the sum is 50, the even division into eight parts results in 6, 6, 6, 6, 6, 6, 7, 7. These values are arranged so that 1/2 of 50 (25) is found on the first four pendants and 1/2 on the last four pendants. Similarly, for the sum 102, the value is evenly divided between the first and last four pendants (51 each), and one set of these is, in turn, divided evenly into four parts (12, 13, 13, 13). Another of the groups has values showing that the division was further restricted to results that are exact multiples of 10. That is, 300 divided into two parts of 150 each, and the 150 subdivided into four parts of 30, 40, 40, 40.

Quipu Example 7.1

The theme of halving pervades quipu example 7.1. The cord arrangement, the values on the cords, and even the way the values are represented are all involved.

Concentrating on the best preserved cord groups that have nine or ten pendants, the values in each group have the following pattern:

$$N, N, N/2, N, N, N, N, N/2, N/2 \qquad \text{nine pendant pattern}$$
$$N, N, N/2, N, N, N, N, N, N/2, N/2 \qquad \text{ten pendant pattern}$$

N or $N/2$ is repeated consecutively 2, 1, 4, 2 times or 2, 1, 5, 2 times. Thus, a number of pendants with value N is followed by half that number of pendants with value $N/2$ (if 2 is recognized as half of 5 when confined to integers). For the different groups, the values of N are 24, 9, 8, 8, 4, 8, 4, 2 with $N/2$ being 4 when N is 9. The last five of these groups are consecutive (with no broken or other size groups intervening). Notice that their values also show a halving pattern: 8, 4, 8, 4, 2.

The number representation on this quipu has the peculiarity that all long knots with value 8 are constructed in two equal parts. The parts are not separated by space and so are not considered as two different knot clusters. Thus, 8 is being represented as 4 + 4. In one group, half of this value is represented by a long knot in two parts, that is, by 2 + 2, and in the others by a single long knot of 4. This is comparable to what we refer to as the distributive axiom. Namely, the results of $\frac{1}{2}(x + x)$ can be arrived at by first adding x to x and then taking half of it

$$\tfrac{1}{2}(x + x) = \tfrac{1}{2}(2x) = x,$$

or by first taking half of each x and then adding

$$\tfrac{1}{2}(x + x) = \tfrac{1}{2}x + \tfrac{1}{2}x.$$

There are also subsidiaries attached to the pendant cords. For almost all of these eight groups, the subsidiaries carry through the halving pattern. In the first group, for example, the pendants with values described by $N = 24$ have two subsidiaries, each with value 2. Then the pendants with values described as $N/2$ have two subsidiaries each of value 1. Therefore, the pattern of the group remains the same but the values are modified to

$$N = 24 + 2 + 2, \qquad N/2 = 12 + 1 + 1.$$

In another group, it is, instead, the number of subsidiaries that is halved. That is, in the group for which $N = 9$ and $N/2 = 4$, the pattern of the group remains the same but the values are modified to

$$N = 9 + 1 + 1, \qquad N/2 = 4 + 1.$$

There is the interesting problem of how to halve a single subsidiary of value 1. The number of subsidiaries cannot be exactly halved, nor can the value on the subsidiary be exactly halved. This problem and the quipumaker's solution occur in two of the groups. The value $N/2$ should appear in three positions in each group. Sometimes a subsidiary of value 1 is present and sometimes it isn't. That is, on this quipu $\frac{1}{2}(1)$ is sometimes 1 and sometimes 0.

Division and multiplication are intimately connected. In describing the relationship between the numbers 8 and 4, the 4 can be viewed as half of 8 or the 8 can be viewed as double 4. We have presented the foregoing quipu examples as division. In quipu example 7.1, we emphasize that it is a halving pattern. In fact, we cannot recast the statement into multiplication. The number pairs on the quipu are (24, 12), (9, 4), (8, 4), (4, 2), (2, 1), and (5, 2). While there would be no problem in saying that 24 is double 12 or that 2 is double 1, there would be no justification for saying that double 4 is 8 *and* double 4 is 9, or that double 2 is 4 *and* double 2 is 5. The fact that division resulting in integers necessitates rounding, justifies calling 4 both half of 8 and half of 9, and 2 both half of 4 and half of 5. In other words, multiplication and division are not inverse operations when the numbers involved are integers.

Quipu Example 7.2

This quipu contains a chart of sixty values, p_{ij} ($i = 1, \ldots, 10; j = 1, \ldots, 6$). The values are repetitions of only four different numbers (11, 12, 20, 21) and the sum of all the values in the chart is exactly 1,000. Forming the categorical sums for $i = 1, \ldots, 10$, we find that their values correspond to simple fractional parts of 1,000. Keeping in mind that only integers are being used, the sums, to within ± 1, are these fractional parts:

$$1/8, \ 3/25, \ 1/8, \ 1/14, \ 1/9, \ 2/25, \ 1/9, \ 1/14, \ 1/9, \ 1/14.$$

Obviously, any ten values adding to 1,000 correspond to fractional parts of 1,000. It is, however, the simplicity of the fractions that captures our attention and leads us to believe that they are the result of conscious division. Furthermore, their accuracy is to within ± 1 although the smallest chart values are elevens. And, simple as these fractions are, they can all be derived from even simpler fractions. Combining the fractions with the same denominators, the whole can be viewed in four parts: $1/3$, $1/4$, $1/5$, and the remainder $3/14$. Therefore, 1,000 could be divided into these parts, and then these in turn divided into smaller parts. Specifically, the ten fractions can be obtained from:

1 divided into $1/3$, $1/4$, $1/5$, $1 - (1/3 + 1/4 + 1/5)$
then, $1/3$ divided into thirds = $1/9$, $1/9$, $1/9$;
$1/4$ divided in half = $1/8$, $1/8$;
$1/5$ divided into fifths and grouped into two parts = $2/25$, $3/25$;
remainder divided into thirds = $1/14$, $1/14$, $1/14$.

We do not intend to imply that this scheme was used, but only to emphasize the simplicity of the fractions. If the ten categorical sums were fractional parts of 1,000 such as $172/627$, $29/217$, $3/14$, $4/111$, and so on, we might still view the distribution of the sixty values in the chart as planned, but we would not believe that the plan was based on division.

5 There are several quipus on which many of the values are related to fundamental units other than one. In the example just cited (quipu example 7.2), all the values in the chart are 11, 12, 20, 21 repeated over and over again. In an earlier example (quipu example 5.1) which has two subcharts, one subchart has twenty-eight values which are repetitions of four numbers (20, 40, 50, 60) and the other has twenty-eight values repeating a different set of four numbers (20, 50, 51, 52). Other quipus, rather than including repetitions of some few numbers, contain many multiples of a few numbers. Where there are many multiples of one or two numbers, and where those numbers are emphasized through some construction feature, we view the numbers as significant to the quipu and the multiples as the result of intentional multiplication. An example of such a quipu follows.

Quipu Example 7.3

The numbers eighteen and seven are physically prominent and the pendant values include many multiples of them. In addition, the multiples are consistently placed within the logical structure of the quipu.

The example is actually a small quipu tied to a slightly larger quipu. Two pendants are set off from the rest by being suspended from a loop at the end of the main cord of the smaller quipu. The small quipu also has two cord groups of three and four pendants respectively. The values on the two specially suspended pendants are eighteen and seven. Each value in the next cord group is a multiple of eighteen or seven. The larger quipu is a chart of twenty-four values, p_{ij} ($i = 1, 2, 3$; $j = 1, \ldots, 8$). Consistently, for all i, for $j = 4$ and $j = 7$, the values are multiples of eighteen or seven. There are a few other multiples elsewhere. Another consistency is that every value greater than fifteen is a multiple of eighteen or seven. In all, with some repeated, the multiples that appear are 18, 36, 54, 72, and 7, 14, 21, 28, 35, 70.

Quipu Example 7.4

This is a quipu on which the number thirteen, and its double twenty-six, dominate the values. The numbers themselves appear on individual pendants or they are the sums of groups of values. What is striking about the quipu is that all the nonzero pendants and their first level subsidiaries are related to these two values. Only three cords, the subsidiaries of the subsidiaries, have unrelated values. All the construction features (cord grouping, cord level, and color coding) reinforce the exhibition of the significant values.

5 (*a*) Quipu example 7.3 is AS63. Quipu example 7.4 is AS151.

(*b*) For an interesting discussion of algebra and rituals, see A. Seidenberg, "The Ritual Origin of Geometry," *Archive for History of Exact Sciences* 1(1962): 488–527.

(*c*) The relationship of calendars and the nineteen-year cycle is briefly discussed in O. Neugebauer, *The Exact Sciences in Antiquity* (New York: Harper Torchbook ed. 1962), pp. 6–8, 102, 140. Specific details about the use of the nineteen-year cycle in constructing a calendar based on the lunar month and solar year can be found in George Zinberg, *Jewish Calendar Mystery Dispelled* (New York: Vantage Press, 1963).

(*d*) Some earlier writers, particularly E. Nordenskiöld, believed that many quipus contained astronomical values. That interpretation ignored the physical and logical structures of the quipus and was based on freely recombining and doing calculations with the numbers on the quipus. Nordenskiöld's astronomical interpretations are in his "The Secret of the Peruvian Quipus," *Comparative Ethnological Studies* 6, pt. 1 (1925): 1–37 and "Calculations With Years and Months in the Peruvian Quipus," *Comparative Ethnological Studies* 6, pt. 2 (1925): 1–35. Our critique of his calculations can be read in our "Code of Ancient Peruvian Knotted Cords (Quipus)," *Nature* 222 (1969): 529–33.

The nonzero pendants are spaced into two groups of six and twelve pendants respectively. Within the groups, subgroups are formed by consecutively colored cords. The array can be partitioned by spacing, color, and cord levels. The values of the individual pendants or the sum of grouped values are shown in figure 7.1 superimposed on a diagram of the partitioned array.

Fig. 7.1. Quipu example 7.4.

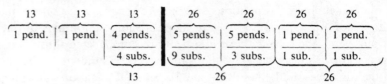

Just as we do not know how sums, fractional parts, or multiples were calculated, we do not know why a quipu emphasizes a particular number. The value could be intrinsic to some situation and thus be recurrent in the data; the quipu could be a demonstration of interest in the number itself; or the number could have significance because it has been invested with some meaning beyond its numerical value. The discussion of structures of individual quipus in the previous chapter focused on patterns expressed through color, cord placement, and number labels. Significant numbers and patterned arithmetic combinations of them could be another facet of such logical forms.

As a last example of a quipu with a physically and arithmetically prominent number, we present one that engages our interest because it combines several ideas that have been discussed. It shows cross and hierarchical categorization as well as multiples of a significant number. The number itself attracts attention because of its association with calendrics.

Quipu Example 6.5—*Continued*

By color and spacing, the quipu consists of three parts. The first part is just one pair of pendants with subsidiaries. The second part is a chart of fifteen elements p_{ij} ($i = 1, 2, 3; j = 1, 2, \ldots, 5$) where each element is a data tree. A schematic of the general data trees for the p_{1j} ($j = 1, 2, \ldots, 5$) and the p_{3j} ($j = 1, 2, \ldots, 5$) were shown in figure 6.12. On figure 6.12 can be seen that each tree has two branches on the first level and that many branches emanate from the first of these and only one branch emanates from the other. Just before the second group of trees (the p_{2j}'s), there is one extra pendant. The third part of the quipu consists of seven pendants which differ from the rest of the quipu in color pattern and in having no subsidiaries. They are preceded by one pendant set off by itself.

Physically, then, there are two pendants that stand out. The value of one is 627 (= 33 × 19) and the value of the other (which precedes the third part of the quipu) is 19. The sum of the seven values in the third part is also 19.

When the values on the data trees are examined, several multiples of nineteen are found. On three consecutive trees, p_{1j} ($j = 1, 2, 3$), the values on level 1 branches are pairs of consecutive numbers (for example, 512 and 513). One value in each pair is a multiple of nineteen (532, 513, 304). Some categorical sums of values on the data trees are also multiples of nineteen. If, for the five p_{1j} ($j = 1, 2, 3, 4, 5$), the values on the second branches of the first level and all the branches that emanate from them are summed, the result is 2,679 (19 × 141). The same sum for the five p_{2j} ($j = 1, 2, 3, 4, 5$) is also a multiple of nineteen (1,805 = 19 × 19 × 5). There are several other individual values and sums that are related to nineteen but they are less consistently placed.

When the first part of the quipu is viewed as a single data tree, it too has multiples of nineteen and values that can be related to the data trees forming the second part of the quipu. The value of one of the second level branches is a multiple of nineteen (38) and the sum of all second level branches is a multiple of nineteen (76). The relationship to the chart of data trees is to the third group p_{3j} ($j = 1, 2, 3, 4, 5$). It involves fractions and, therefore, exhibits another of the ideas that have been discussed. Recall that the five trees in the third group and the tree of part one have only two branches on level 1. When the categorical sums for the third group are formed for each of these two branches, each sum is the same fractional part of the value on the corresponding branch on the tree in part one. For each, to within the limitation of integers (±1), the fractional part is the simple fraction 1/10.

Figure 7.2 is a capsulization of the observations about this quipu.

Fig. 7.2. Quipu example 6.5. Each p_{ij} is a data tree; p_{1j} and p_{3j} are shown in figure 6.12.

Quipu Layout					Some Relationships
1 tree					Sum of level 2 branches = 19 × 4
p_{11}	p_{12}	p_{13}	p_{14}	p_{15}	Multiples of 19 on level 1 branches of p_{11}, p_{12}, p_{13}, Sum of level 1 second branches and their branches = 19 × 141
p_{21}	p_{22}	p_{23}	p_{24}	p_{25}	Sum of level 1 second branches and their branches = 19 × 19 × 5
p_{31}	p_{32}	p_{33}	p_{34}	p_{35}	Sum of level 1 branches each = $^1/_{10}$ level 1 branches on top tree
7 pendants					Sum = 19

627

19

The number nineteen is of calendric concern because it is related to the alignment of cycles based on the sun with cycles based on the moon. In a calendar based on the appearances of the moon, months are 29 or 30 days so that the same phases of the moon occur on more or less the same date each month. In a calendar based on the earth's revolution about the sun, a year is 365 or 366 days so that the equinoxes and solstices occur on more or less the same date each year. If both relationships are considered, the time it takes to return to the same association of moon, earth, and sun is nineteen solar years or 235 lunar months. To retain some synchronism between a lunar calendar and the seasons, lunar months are combined into lunar years. But the lunar years have to vary, with some having 12 months and others 13. If there are basically 12 lunar months in a year, then within the nineteen-year cycle, 7 more months have to be inserted. Similarly, the nineteen-year cycle would be important to those with a solar calendar if they wanted to calculate the dates of new moons.

A quipu emphasizing the number nineteen, containing multiples of nineteen and categorical sums which are multiples of nineteen, raises the possibility of a calendric association. The seven final pendants summing to nineteen are suggestive of a scheme that divides the nineteen-year cycle into seven unequal parts in order to place the seven additional months.

As intriguing as the idea sounds, there would have to be a considerable amount of associated evidence before any single quipu could be so specifically interpreted. Nevertheless, it is certainly plausible that some quipus are recordings of astronomical values or other fixed values related to the natural environment.

6 Before discussing numerical relationships that demonstrate more intricate arithmetical ideas, a final point about number representation needs to be made. We have distinguished between the idea of fractional values and a symbolic representation of them. We distinguish further between a variety of representations and a standardized representation that is generally used.

The use of spaced knot clusters with particular types of knots to represent integers on quipus is sufficiently standardized to be recognized by us. Moreover, they occur on some quipus where the interpretation can be corroborated. If different quipumakers had the same basic idea but represented it more individualistically, we possibly would not be able to recognize it. The history of Western notation for fractions and decimal fractions is an interesting example of varied and individualistic expressions for the same basic idea. It serves as a reminder that unifor-

6 Notations for fractions are discussed in Cajori, *History of Mathematical Notations*, pp. 309–14. Notations for decimal fractions are discussed on pp. 314–35. The idea of decimal fractions and their notations are also discussed in George Sarton, "The First Explanation of Decimal Fractions and Measures (1585). Together with a History of the Decimal Idea and a Facsimile (no. XVII) of Stevin's *Disme*," *ISIS* 23 (1935): 153–224. See in particular pt. III (secs. 80–101), pp. 167–86 of the article.

mity of number representation is more common to us in our age of standardization than it was in other times or places. As late as the sixteenth century, individuals used special signs for special fractions, such as ~ or $\doteq$ (in England) and $\overset{o}{=}$ or $\overset{o}{m}$ (in Spain) for one-half, and • for one-fourth and $\check{\tau}$ for three-fourths (in England). A horizontal line separating the numerator and denominator (i.e., $\frac{3}{4}$) became general usage only in the sixteenth century, but in the seventeenth century there were still those who omitted it. Later, starting about 1850, because fractions had to be compressed into a single line of print, many people began to use the solidus instead (i.e., 3/4). Now, as common practice, we use both forms. Similarly, the idea and clear statement of decimal fractions is attributed to Simon Stevin (Belgium, 1585) but our notation for them did not become standardized for hundreds of years. Stevin wrote

$$\overset{⓪①②③④}{23759}$$

for the value we symbolize as 2.3759. Examples of other individuals' notations are:

$$\overset{o}{12}\overset{iii}{3}.45\overset{vi}{9}.872 = 123.459872 \quad \text{(Beyer, Frankfurt, 1603)}$$

$$693② = 6.93 \qquad \text{(von Kalcheim, Bremen, 1629)}$$

$$16\,|\,\overset{o\ o\ o\ o}{7249} = 16.7249 \qquad \text{(Jager, London, 1651)}$$

$$3:04 = 3.04 \qquad \text{(Rawlyns, England, 1656)}$$

$$31.\underline{008} = 31.008 \qquad \text{(Foster, London, 1659)}$$

$$22=3 = 22.3 \qquad \text{(Caramuel, Bohemia, 1670)}$$

$$732,|\underline{569} = 732.569 \qquad \text{(Raphson, London, 1728)}$$

$$4.2\overset{i\ iv}{5} = 4.2005 \qquad \text{(Chelucci, Rome, 1738)}$$

And, of course, standard usage today differs from France (4,5) to England (4·5) to the United States (4.5).

7 So far, in the quipu examples, when referring to fractional parts, we have used fractions rather than decimal fractions. Since we will now begin to use decimal fractions, we pause to first explain why and to explain the difference in the associated statements about accuracy.

We have used statements of the form: "to within ± 1, $X = \frac{1}{8} Y$." In each case, we focused on an integer (X), and associated with it another integer (Y) and a fraction that related them. The accuracy statement "to within ± 1" was connected to the integer X. In the examples that follow, we wish to compare ratios of numbers. For example, we want to be able to say that $a/b = c/d$. Translating a/b and c/d into decimal fractions takes the attention off the specific integers involved and places it on the result of combining them. Similarly, our accuracy statement must be about the values a/b and c/d rather than about any specific integer.

Consider several fractions that are near 50/100 or 100/200: 49/99, 50/101, 51/101, 99/200, 101/200, 99/199, 100/199, 100/201, 101/201. Both the numerators and the denominators vary, but they vary together. The numerators from 49 to 51 are combined with denominators from 99 to 101 and those from 99 to 101 are combined with denominators from 199 to 201. It is easier to see how much, if at all, the variation effects the results when decimal fractions are used instead. If each of these fractions is translated into a decimal fraction that is rounded to three positions, they are all near 0.500. All the values are between 0.495 and 0.505 and so the maximum deviation from 0.500 is -0.005 or $+0.005$. Since 0.005 is 1 percent of 0.500, we say that the fractions all have the value 0.500 to within 1 percent. Fractions with quite different numerators and denominators could be included in this generalization. For example, 76/151, 106/213, and 1107/2197 all have the value 0.500 to within 1 percent.

8 Quipu example 7.5 is AS120. Quipu example 7.6 is AS143.

8 Quipu Example 7.5

This quipu consists of a chart of twenty-four values p_{ij} ($i = 1, 2, 3$; $j = 1, \ldots, 8$) and the categorical sums for $j = 1, \ldots, 8$. Also, for each i, for $j = 3$, there is a subsidiary value. Their categorical sum is on a subsidiary attached to the $j = 3$ categorical sum. To refer more easily to the categorical sums, we call them p_{sj} ($j = 1, 2, 3, 3$ subsidiary, $4, \ldots, 8$); the value of each p_{sj} is $p_{1j} + p_{2j} + p_{3j}$. No two values in the chart are the same and the range of values in the chart is considerable; the smallest is 102 and the largest is more than 400 times as much, namely 43,372. But, the ratios of the values in the chart to their categorical sums are remarkably consistent. The numerators of the nine ratios p_{1j}/p_{sj} ($j = 1, 2, 3, 3$ subsidiary, $4, \ldots, 8$) range from 102 to over 140 times as much (14,743), and the denominators range from 300 to 43,372. But, for all but $j = 3$, each ratio is 0.340 to within 0.6 percent. (The exception, $j = 3$, deviates from 0.340 by 1.4 percent.) Similarly, for the nine ratios p_{2j}/p_{sj}, with the exception of $j = 1$, each is 0.425 to within 0.7 percent. And to within 0.9 percent (again with the exception of $j = 1$), each p_{3j}/p_{sj} is 0.235. Comparing the chart values to each other, there is a simple fraction relating each of the nine p_{1j} to the corresponding p_{2j} and p_{3j}. To within 1 percent, each p_{1j} is 17/33 of $p_{2j} + p_{3j}$. ($j = 3$ is still an exception; it deviates by 2.2 percent.)

Quipu Example 7.6

Consistency of ratios is also found on quipu example 7.6. It too has a large range of values. In fact, it contains the largest number we have seen on any quipu, 97,357. Even more interesting is that the ratios themselves are similar to those in the previous example.

Part of the quipu is a chart of sixteen values p_{ij} ($i = 1, \ldots, 4$; $j = 1, \ldots, 4$) and the categorical sums for $j = 1, \ldots, 4$. The values in the chart range from 2,026 to 22,469 with no two being the same. Again, calling the categorical sums p_{sj} ($j = 1, 2, 3, 4$) and finding the ratio of each p_{ij} to its categorical sum: $p_{1j}/p_{sj} = 0.110$; $p_{2j}/p_{sj} = 0.228$; $p_{3j}/p_{sj} = 0.437$; and $p_{4j}/p_{sj} = 0.225$. Each of these is within 1.6 percent with the exception of $j = 4$ for $i = 3, 4$. (Curiously p_{44}/p_{s4} deviates by exactly 10 percent.) Comparing the chart values to each other, the same simple fraction as appeared in the last quipu relates the p_{1j} and p_{2j} to the corresponding p_{3j} and p_{4j}. To within 1.2 percent (but 2.7 percent for $j = 3$), each sum $p_{1j} + p_{2j}$ is 17/33 of $p_{3j} + p_{4j}$.

Quipu Example 5.3—*Continued*

This is one more quipu that is similar in that it has consistent ratios. It has already been described in detail as an example of a chart which contains categorical sums of values in subcharts. And five values in the subcharts are themselves categorical sums of values in sub-subcharts. The quipu was further elaborated upon when discussing regularities on charts since the values in the chart, when compared to the sums of their respective categories, have consistent ratios. Again, the values in the chart are quite dissimilar as they range from 660 to 21,243. Figure 5.16 shows the ratios of the values to the sums of their respective categories. Now comparing them, we can say that, within 0.8 percent, the ratios (r_j; $j = 1, \ldots, 5$) are $r_1 = 0.122$; $r_2 = 0.017$; $r_3 = 0.534$; $r_4 = 0.105$; $r_5 = 0.222$ for both values of i.

In all three of these quipus, the values have a very large range. Yet, they are the same fractional parts of their own categories and not simple fractional parts of those categories. To have, as in quipu example 7.5, nine values that each are 0.425 of such diverse numbers as 1,068, 5,485, and 42,760, requires a concept of proportions and a scheme to somehow calculate them. If there were evidence of decimal fractions on the quipu, it could just be said that somewhere 0.425 was multipled by each of these numbers. In the absence of evidence of the representation of decimal fractions and of fractions, the conclusion has to be that the Incas clearly could work with fractions without having any representation for them.

9 (a) The Greek study of proportions started in Pythagoras' time (ca. 500 B.C.E.) with the study of the arithmetic, geometric, and harmonic means. The seventh proportion is said to have been added by Myonides and Euphranor and also discussed by Nicomachus (ca. 100 C.E.). For a fuller discussion, see Thomas Heath, *A History of Greek Mathematics,* vol. I (London: Oxford University Press: 1921), pp. 85–86 or M. Ghyka, *The Geometry of Art and Life* (New York: Dover Publications, 1977).

(b) For quipu example 7.5, the value of X is 0.447 and $1-X$ and $1-X^2$ are exact to three figures. For quipu example 7.6, the value of X is 0.482 and $1-X$ and $1-X^2$ are accurate to within 0.7 percent. For quipu example 5.3, X is 0.815 and $1-X$ and $1-X^2$ are accurate to within 0.6 percent. Quite unexpectedly, the values of X^2 for example 7.5 and example 5.3 are the simple fractions 1/5 and 2/3 respectively. When $1-X^2$ in quipu example 7.6 is separated into two parts, $1-X$ and $X(1-X)$, each is accurate to within 0.8 percent. When the unit in quipu example 5.3 is to be split into its three parts, the relative sizes of the three parts require carrying more significant figures. Therefore, the data for $i=1$ was used by itself and all the values were recalculated. Now $X=0.8165$ and the parts $1-X$, X, and $\pi\left(\dfrac{1-x}{2}\right)^2$ are accurate to within 0.4 percent, $1-X^2$ is to within 1 percent and $1-X-\pi\left(\dfrac{1-x}{2}\right)^2$ to within 1.7 percent.

(c) Figure 7.8 is quite similar to a geometric form thought to be important and persistent in the cosmology of western South America. See W. H. Isbell, "Cosmological Order Expressed in Prehistoric Ceremonial Centers," *Actes du XLIIe Congrès International des Américanistes* 4 (1976): 269–97.

9 We now pursue further the values that have appeared as consistent ratios. This is done in order to see if there is anything these values have in common that might give some clues as to the type of calculations they arose from or to their use. We note, first of all, that there is insufficient data available about the three quipus to either confirm or deny that they are related to one quipumaker. They do, however, have the same general provenance and there are similarities in color.

Since there are three values characteristic of quipu example 7.5, a classification scheme devised by the Greeks can be used. They established that for three values, a, b, c such that $a < b < c$, there can be ten different relationships. The seventh relationship,

$$\frac{c-a}{b-a} = \frac{c}{a}$$

applies to this quipu's values. That is, for $a = 0.235$, $b = 0.340$, and $c = 0.425$, $(c-a)/(b-a)$ calculated to three figures exactly equals c/a calculated to three figures.

To view quipu example 7.6 and quipu example 5.3 within the same classification scheme, their four and five characteristic values are reduced to three values each. To reduce them, two or three values are combined together, but without rearranging their order. The three sets of quipu values are summarized in figure 7.3. Note that for each quipu, a, b, c such that $a < b < c$, appear in the same order. Again we find that of the ten different possible relationships, it is the same relationship that applies to quipu example 7.6 and to quipu example 5.3. (The former satisfies the relationship to within 1.6 percent and the latter to within 0.4 percent.) Therefore, the three sets of values, when viewed as having three values each, do indeed have something in common.

Fig. 7.3. Consistent ratios.

	Quipu Example 7.5	Quipu Example 7.6	Quipu Example 5.3
b	0.340	0.110 $\Big\}$ 0.338 0.228	0.222
c	0.425	0.437	0.105 0.534 $\Big\}$ 0.656 0.017
a	0.235	0.225	0.122

We pursue their commonality further in an attempt to better visualize it and to relate to it the individual values that were combined together. We present a spatial visualization that also serves as a basis for a simple scheme for generating the values.

Consider, first of all, any three values $a < b < c$. Define c as a unit and a and b as parts of that unit. Associate c with the area of a unit square and a with the area of a rectangle formed by cutting a strip off a unit square. Algebraically, the values a, b, c are being normalized to a/c, b/c, and 1 and then a/c is being relabeled $1-X$.

Fig. 7.4

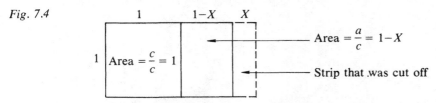

We next seek to interpret b/c as the area of a figure related to the unit square and the rectangle. Because of the specific relationship that exists between a, b, and c, $b/c = 1-X^2$. That corresponds to cutting from a unit square a smaller square whose length and width are each the width of the strip that was cut off before. Figure 7.5, therefore, represents the commonalities of the three quipu ratios. The ratios of the areas of parts of this figure to the area of the whole figure are the characteristic ratios of the quipus. For each quipu there is a different value of X involved.

Fig. 7.5

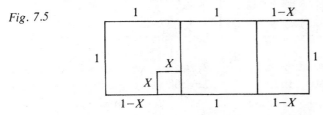

For quipu example 7.6, the value designated as b was the sum of two values. Splitting b into its two original parts, we find that the parts neatly fit into our picture. The portion of the figure associated with b simply needs to be separated along the line that was used to define the width of the small square.

Fig. 7.6

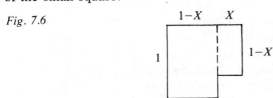

For quipu example 5.3, the value designated as c was the sum of three values so, in the picture, it is the unit square that must be partitioned. Again we find that the partitioning depends on X in a straightforward way. The three values separate the unit square into a rectangle of width X, a circle of diameter $1-X$, and a rectangle of width $1-X$ from which the circle has been subtracted.

Fig. 7.7

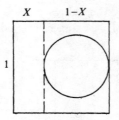

The three sets of ratios, with their three, four, or five values, are, therefore, still summarized by the original figure. Now, however, the figure has more subdivisions within it (see fig. 7.8).

Fig. 7.8

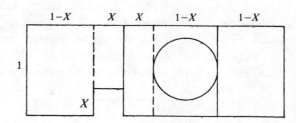

The model presented in figure 7.8 can be restated as a numerical scheme for generating the ratios in question. To generate a set of three values:

1. Take some whole and divide it into three equal parts: P, P, P.
2. Leave one part as is: P
3. Withhold some fraction of the second part: $P(1-X)$
4. Withhold the square of that fraction of the third part: $P(1-X^2)$.

Let us, as an example, generate some numbers according to this scheme. Starting with 3,000, divide it into three equal parts of 1,000 each. Leave one part as is (1,000); delete, say 1/3, of the second part (1,000 − 333 = 667); delete $(1/3)^2$ = (1/9) of the third part (1,000 − 111 = 889). The resulting numbers are 1,000, 667, and 889 and their sum is 2,556. The ratios of the numbers to their sum are 0.391, 0.261, and 0.348. Starting with a different whole, say 4,230, and using the same fraction 1/3, the resulting numbers are 1,410, 940, and 1,253. Their sum is 3,603 and again the ratios of the numbers to their sum are 0.391, 0.261, and 0.348. Calling the ratios a, b, c such that $a < b < c$, they satisfy $(c-a)/(b-a) = c/a$. Following the same scheme but using a fraction other than 1/3, would lead to different ratios. Nevertheless, the ratios would still satisfy $(c-a)/(b-a) = c/a$.

For the further subdivision needed in quipu example 7.6, the third number that was generated is split into two parts; one equal to $P(1-X)$ and the other to whatever remains. For the numbers of our first example, 889 is split into 667 and 222. For the further subdivision needed in quipu example 5.3, the first number (P) is separated into PX and $P(1-X)$, and then $P(1-X)$ is separated into two values in the ratio of $4-\pi$ $(1-X)$ to π $(1-X)$. In the example, 1,000 is separated into 333 and 667. Then 667 is separated into 349 and 318.

In sum, our conclusions about these three quipus are:

1. On the individual quipus, numbers are consistent fractional parts of their categories indicating that they were the result of calculation.

2. While the fractions are different for the three quipus, they can result from the same general calculating scheme. A spatial visualization and a numerical scheme were introduced to illustrate the relationship of the three sets of values.

10 Quipu Example 7.7

The last quipu example is, arithmetically, the most interesting. In all, it contains only thirteen values, but their interrelationships are intricate and can be expressed in many different ways. Physically, it is two small quipus that were found together. One quipu has seven pendants and three subsidiaries and the other has just three pendants. Ignoring, for the moment, the first pendant and the subsidiaries, the quipu form a chart of values p_{ij} ($i = 1, 2, 3$; $j = 1, 2, 3$) where values for $i = 1, 2$ are on the larger quipu and those for $i = 3$ are on the smaller one. In tabular form, the chart is:

Fig. 7.9

p_{11}	p_{12}	p_{13}
p_{21}	p_{22}	p_{23}
p_{31}	p_{32}	p_{33}

The values in the chart range from 734 to 2,427.

10 Quipu example 7.7 is AS55 and AS56. We have modified one digit of one number on AS55 by one. As noted before, we do not consider an error of one in one knot cluster to be very significant, since errors in knotting or our errors in counting knots or in transcription are all possible.

One attractive relationship is that the product of the values in the third row is the geometric mean of the products of the first and second rows. That is,

$$\frac{p_{11}\,p_{12}\,p_{13}}{p_{31}\,p_{32}\,p_{33}} = \frac{p_{31}\,p_{32}\,p_{33}}{p_{21}\,p_{22}\,p_{23}}$$

And, the product of the values in the third row is also the same as the product of the diagonal values:

$$p_{31}\;p_{32}\;p_{33} = p_{11}\,p_{22}\,p_{33}.$$

When comparing the individual values to each other, three simple fractions keep reappearing. They are $\frac{11}{14}, \frac{7}{8}$, and $\frac{34}{33}$. The values in the third row, when multipled by these fractions, result in the values in the second row. And, the values in the first row, when multipled by them in a different order $\left(\frac{7}{8}, \frac{34}{33}, \frac{11}{14}\right)$, result in the third row. This cyclical relationship can be summarized by:

$$\frac{p_{2j}}{p_{3j}} = \frac{p_{3,j+2}}{p_{1,j+2}} \text{ for } j = 1, 2, 3 \text{ where addition is mod 3.}$$

The table of values can, therefore, be rewritten in terms of the fractions and only the original values in the first row:

Fig. 7.10

	p_{11}			p_{12}			p_{13}
$\left(\dfrac{11}{14}\right)\left(\dfrac{7}{8}\right) p_{11}$			$\left(\dfrac{7}{8}\right)\left(\dfrac{34}{33}\right) p_{12}$			$\left(\dfrac{34}{33}\right)\left(\dfrac{11}{14}\right) p_{13}$	
$\left(\dfrac{7}{8}\right) p_{11}$			$\left(\dfrac{34}{33}\right) p_{12}$			$\left(\dfrac{11}{14}\right) p_{13}$	

Closer scrutiny of the values in the first row shows that they, too, are related to each other by these fractions. Specifically,

$$p_{11} = \left(\frac{11}{14}\right)\left(\frac{7}{8}\right) p_{12}.$$

(All of the foregoing statements are to within 0.4 percent. What is more, ten out of the twelve of them deviate by at most only 0.2 percent.)

So far, we have ignored the first pendant value. Having observed, as a construction feature, that sums are frequently on individual pendants set off at the end of the main cord, we examine it to see if it is the sum of values in the chart. We find, still within 0.2 percent, that it is the sum of the values in the first row divided by one of the omnipresent fractions. Its value is $(p_{11} + p_{12} + p_{13})/(34/33)$. The pendant has two subsidiaries (s1, s2) and there is also a subsidiary (s3) on the pendant with value p_{12}. The value of s1 is related to the value of s3 through the fractions. To within 0.5 percent, $s1 = (\frac{7}{8})^2 (s3)$ or to within 0.9 percent, $s1 = \frac{11}{14} (s3)/(34/33)$. The fact that this last relationship can be expressed in these two different ways raises the unusual idea that the three dissimilar looking fractions could themselves be related. Actually, $\frac{7}{8}$ is a very good approximation to $\sqrt{(11/14)/(34/33)}$. The approximation is accurate to 0.2 percent. Because p_{11} can be expressed in terms of p_{12} and using the approximation $(\frac{7}{8})^2 = \frac{11}{14}(\frac{33}{34})$, the nine table values can be reduced to dependence on only two of the original values and two of the fractions. To simplify their presentation, let $B = \frac{7}{8}$, $C = \frac{34}{33}$, $x = p_{12}$, and $y = p_{13}$. The table then has the form:

Fig. 7.11

CB^3x	x	y
C^2B^6x	BCx	C^2B^2y
CB^4x	Cx	CB^2y

All the relationships already described can be seen from this formulation. For example, both the product of the values on the diagonal and the product of the values in the third row are $C^3B^6x^2y$. And, of course, additional ones can now be found. For example, the products of the off-diagonal values p_{12}, p_{23}, p_{31} and of p_{13}, p_{21}, p_{32} are the same as the product of the diagonal values.

The values on this quipu pair must have resulted from intentional calculations. The interrelationship of the values depends on fractions and on logic that is more complex than is used to yield values that are consistent fractional parts of their whole. But, if the two small quipus were not together, it is unlikely that any of the relationships would be seen.

11 Taken together, the individual quipus demonstrate that the body of arithmetic ideas used by the Incas must have included, at a minimum, addition, division into equal parts, division into simple unequal fraction-

al parts, division into proportional parts, multiplication of integers by integers, and multiplication of integers by fractions. These ideas are often seen to be embedded within the logic of cross and hierarchical categorization.

In ending the discussion of individual quipus, we note that four of the quipus used as examples consisted of two independent parts that are physically related. In quipu example 6.8, a quipu with an *aba* pattern is linked to a quipu with an *aaba* pattern. In quipu example 6.9, a long tail attached to a quipu contains a restatement of the quipu's theme. In quipu example 7.3, the numbers eighteen and seven are physically prominent on one quipu and consistently placed multiples of them are found on an attached quipu. In the last example, the three values on one quipu complement the values on the other quipu so that cyclical patterns are formed. We cannot know whether these were made individually and then joined because they were related, or whether it was intended that they eventually be separated. Either case emphasizes that quipus have a conceptual context that extends beyond any single quipu.

EXERCISES

Exercise 7.1 The quipus are no. 3 and no. 5 in Carlos Radicati di Primeglio, "La 'Seriación' como Posible Clave para Descifrar los Quipus Extranumerales," *Documenta: Revista de La Sociedad Peruana de Historia* 4 (1965): 112–215. For purposes of the exercise, we have modified two digits in one chart and one digit in the other.

Exercise 7.1
The following charts are excerpted from a pair of related quipus. Find numerical patterns within each chart that are consistent for both sets of values.

Fig. 7.12

1	7	4	4	4
1	11	1	4	4
1	12	2	4	5
1	14	2	5	6
1	16	2	6	2

6	4	4	10	10
6	8	6	10	12
6	14	8	12	17
6	22	12	17	14
6	34	15	14	15

An Answer to Exercise 7.1
Each chart has twenty-five values p_{ij} ($i = 1, \ldots, 5$; $j = 1, \ldots, 5$). Within each chart:

1. p_{i1} is the same for all i;
2. $p_{i5} = p_{i+1,4}$ for $i = 1, 2, 3, 4$;
3. $p_{i2} + p_{i3} = p_{i+1,2}$ for $i = 1, 2, 3, 4$;
4. $p_{14} = p_{15}$; $p_{53} = p_{55}$.

Exercise 7.2

1. Complete the table below by following the numerical scheme described in section 9 of this chapter. Start with the numbers 300 and 1,000 and use the fraction 1/2. All results must be stated as integers.

Fig. 7.13

Sum	2,250	375		
P	1,000	166		
$P(1 - X)$	500	84		
$P(1 - X^2)$	750	125		

2. Calculate the ratio of each value in the second row to the corresponding sum in the first row. Then find a decimal fraction that best describes the four results.

3. Do the same for the third row.

4. Do the same for the fourth row.

5. For the three values obtained from 2, 3, and 4, call the smallest a, the next b, and the largest c. Verify that $(c-a)/(b-a) = c/a$.

6. Calculate the ratio of each value in the second row to the sum of the corresponding values in the third and fourth rows. Find a simple fraction that describes the four results.

An Answer to Exercise 7.2

1. The numbers that result when starting with 300 are 100, 50, and 75. Their sum is 225. Starting with 1,000, the results are 333, 167, and 250 with a sum of 750. Because 1,000 is not evenly divisible by three, and 333 is not evenly divisible by two, the results might differ by 1.

2. The ratios are 0.444, 0.443, 0.444, 0.444. The best description is 0.444.

3. The ratios are 0.222, 0.224, 0.222, 0.223. The best description is 0.223.

4. The ratios and best description all are 0.333.

5. For $a = 0.223$, $b = 0.333$, $c = 0.444$; $(c-a)/(b-a)$ and c/a are each 2.00 ± 0.01. (That is to within 0.5 percent.)

6. The ratios are 0.800, 0.794, 0.800, 0.799. Since 0.800 equals 4/5, the maximum deviation from 4/5 is 0.006. All can be described as 4/5 to within 0.8 percent. It is important to note that all the deviations in parts 2 to 6 are due to the restriction to integer results in part 1.

Exercise 7.3 The Treviso book is discussed in D. E. Smith and J. Ginsburg, "From Numbers to Numerals and from Numerals to Computation," in *The World of Mathematics,* ed. J. R. Newman, vol. 1 (New York: Simon & Schuster, 1956): pp. 442–64. The grating method is also discussed in Florian Cajori, "The Hindus" in *The Growth of Mathematics,* ed. R. W. Marks (New York: Bantam Books, 1964), and in Al-Daffa', Ali Abdullah, *The Muslim Contribution to Mathematics* (London: Croom Helm Ltd., 1977).

Exercise 7.3

Multiplication using the "grating method" was taught in fifteenth-century Europe. According to some historians, it was used earlier in India and still earlier by the Arabs. This is how multiplication looked in the earliest printed arithmetic (anon., Treviso, Italy, 1478).

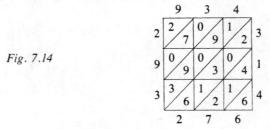

Fig. 7.14

In the illustration, 934 is being multiplied by 314 so 934 is written across the top and 314 down the side. Each position in the chart is filled in with the product of the digit at the top of the column and the digit at the side of the row. Thus, row 1 contains the results of 9 × 3, 3 × 3, and 4 × 3. Notice that 9 × 3 equals 2/7, 3 × 3 equals 0/9, etc. Then, starting at the lower right corner of the chart, add down the diagonal until the bottom of the chart is reached. Record the result below the diagonal. For example, the first diagonal contains just 6 and so the sum 6 is placed below. The second diagonal contains 4, 1, 2 so the sum 7 is placed below. The next diagonal contains 2, 0, 3, 1, 6 so the sum is 12. A 2 is placed below and the 1 is carried to be added in with the next diagonal. The answer 293,276 is read starting at the upper left.

1. Calculate 413 × 729 using the grating method.

2. Design a quipu that corresponds to the figure in the Treviso arithmetic. Give the answer in the form of a schematic. Explain the correspondence.

3. Assuming that you had no information other than your schematic, do you think that you could interpret the quipu correctly?

An Answer to Exercise 7.3

1.

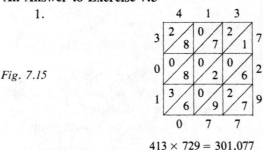

Fig. 7.15

413 × 729 = 301,077

2. Color C1 corresponds to the multipliers. The first pendant contains 934 and the digits 3, 1, 4 are on the last pendant of each group. The groups correspond to the rows in the chart; the positions within the groups correspond to the columns; and the colors C2–C7 correspond to the diagonals. The last group contains the digits of the product.

Fig. 7.16

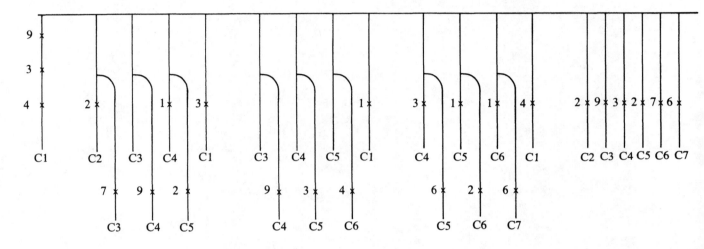

Chapter **8** | A Piece of String

1 The Spanish who heard about quipus in the early part of the sixteenth century discussed them with limited understanding; later in that century, they burned them in the belief that they were ungodly. At the turn of the twentieth century, still other Westerners began to remove quipus from graves. Some of these got into the hands of people who collect or sell the remnants of dead civilizations; some came to rest in the storage drawers of museums. A visually attractive specimen in a museum was sometimes taken out of a drawer and hung behind glass in an exhibit hall.

The pieces of string now began to attract the attention of curious people. One of them was L. Leland Locke who asked for, and received permission to study the quipus in the collection of the American Museum of Natural History in New York. Locke began his studies in 1910–11. In 1923, he recalled his work during the initial year; he writes ". . . one specimen was found which, by the peculiar arrangement of the strands, provided an indisputable key to the significance of the quipu knots." The specimen Locke refers to is a quipu of six groups of four pendants with a top cord on each group. The key was the discovery that the value on the top cord in each group was the sum of the values on pendants in that group when the knots were interpreted as base 10 positional numbers. Locke applied the key to the other quipus in the museum's collection, whether or not they had top cords. In this way, he was led to conclude that all the quipus he studied recorded quantities.

With Locke's achievement, the code of the quipu was generally thought resolved. This is unequivocally stated in Charles W. Mead's brief foreword to Locke's major work. Mead writes: "The mystery has been dispelled and we now know the quipu for just what it was in prehistoric times, and what it is, in its limited use today, simply an instrument for recording numbers." In the middle fifty years of the twentieth century, about forty-five more quipus that had reached museum drawers were described in sixteen separate sources written in Spanish, English, German, French, and Italian, all of them following Locke's lead. But a few people were uneasy with the notion that the principal expressive mode of a major civilization could be resolved in such simple terms. More than thirty-five years ago, John R. Swanton wrote that the quipu had to be ". . . a more highly developed and a much more perfect medium of expression than some recent students have thought . . ."; this, he argued, was ". . . demanded by the very real splendor of the Andean civilization as a whole."

It is possible to think of the quipu as the quintessence of the particular insistence that defined the Inca state, culture, and civilization. With

1 (*a*) The destruction of quipus was ordered by the Council of Lima, 1583. For the history of the church during this period, see Rubén Vargas Ugarte, *Historia de La Iglesia en El Perú*, vol. 2 (1570–1640) (Burgos, Spain: Aldecoa, 1959). The earliest date we can find for the museum acquisition of a quipu is 1895. It is AS122, at the Museum für Völkerkunde, West Berlin.

(*b*) Quipus do not exist solely as museum specimens; they are also found overlayed with contemporary culture. A colorful Peruvian postage stamp issued in 1972 testifies to the continuing interest in quipus by those for whom it is a part of their cultural tradition. Over a doorway on a street in Ica, Peru (655 Avenue José Matias Manzanilla) hangs a sign announcing the office of the Dirección General de Contribuciones—Ministerio de Economía y Finanzas. The figures on the sign are a pan balance and a quipu, thus associating the quipu with the state's taxation function. But, just as a Grecian urn, seen in a museum, inspired Keats's ode, so the quipu has entered the vocabulary of other cultural traditions through the artistic imagination. In 1970, we encountered a sign for the Quipu Basement Theatre on a street in London (49 Greek St., W1). In a letter from Walter Hall we learned that the word was used to signify ". . . an alliance of individuals, knots drawn together by mutual interests, threads. A network of artistic interaction." An American science fiction novel *Babel-17* (Samuel R. Delany, [New York: Ace Books, 1966]) explores the cultural effects of people using computer language rather than natural language. In it, a character explains that a quipucamayocuna [*sic*] is "the guy who reads all the orders as they come through and interprets them and hands them out. . . . They gave orders by tying knots in rope, we use punch cards" (p. 172). We expanded on the quipumaker as a person of privilege in the Inca bureaucracy; in the novel those knowledgable about computers are persons of privilege. Quipus are central in another science fiction novel, *The Martian Inca* by the British writer Ian Watson (New York: Charles Scribner's Sons, 1977). An unmanned spacecraft carrying soil samples from Mars crashes in a Bolivian village. As a result of being contaminated by a Martian chemical, a young villager begins to see visions and, claiming to be the Inca, attempts to put back an Inca-like state. To him, the quipus in museums are symbols of oppression of his people, the quipus in his visions are symbols of unity and enlightenment, and the reintroduction of the use of quipus is one of the specific goals of his revolution. A highly trained specialist in planetary studies, landed on Mars and contaminated by the same chemical, sees visions described in ultramodern scientific terminology. He explains that he is seeing "topological models of filtered reality . . . in n-space" and the Bolivian is seeing quipus because each, in his own imagery, is seeing "the geometry underlying the world" due to the chemical which is "the perfect patterning catalyst" (pp. 167, 168, 182). It is the logical structure of quipus and the quipu as a metaphor for the Inca state that has captured this author's imagination and is being stated in the novel in a way so different from ours.

(c) Leland L. Locke's major work is *The Ancient Quipu or Peruvian Knot Record* (New York: The American Museum of Natural History, 1923). The complete bibliography of quipu descriptions is in our *Code of the Quipu Databook*. For the comments of John R. Swanton, see *The Quipu and Peruvian Civilization*, Smithsonian Institution, Bureau of American Ethnology, Anthropological Papers, no. 26 (Washington, D.C., 1943), pp. 589–96. A similar uneasiness with the Lockean solution is expressed by Gustavo Valcárcel in *Perú. Mural de un Pueblo: Apuntes Marxistas sobre el Perú Prehispánico*, (Lima: Editoria Perú Nuevo, 1965).

(d) The major contemporary theoretician concerned with finding the principles of logic that underlie cultural systems is Claude Lévi-Strauss. His most accessible work is probably *The Savage Mind* (Chicago: University of Chicago Press, 1968).

2 For discussion of the use of knotted strings in other cultures see, for example, Kaj Birket-Smith, "The Circumpacific Distribution of Knot Records," *Folk*, 1966–67, pp. 15–24, and Solomon Gandz, "The Knot in Hebrew Literature, or from the Knot to the Alpha-

pieces of string, the Incas developed a form of recording that forces reconsideration of writing as we generally understand that term. Further, the quipu is an achievement which now must be fit with ideational endeavor, that most crucial part of human history. People all over the world, regardless of technological attainment, have developed intellectual systems which are expressed through such things as myths, social organization, and classification of the natural world. Students of such systems, contending that the same principles of logic underlie all of them, have set out to find what those principles might be. In some cases, their results have been couched in mathematical-like formulations. In theory, and with effort, we can make some limited sense out of the quipu because we share with the quipumaker some logical concepts that cut across space, time, and cultural differences. (That, in fact, is what we were doing in the preceding three chapters.) Ideas about quipus, however, are more amenable to mathematical expression than, say, myths or social organization, because the quipu is itself a numerical-logical system expressed in tangible spatial configurations.

2 As a numerical-logical system, quipus are related to that part of the human intellectual endeavor generally regarded as mathematics. Since the early statement by Locke that quipus were number records, they have been given some cursory attention in mathematical literature. In some texts, quipus are in that first chapter which refers to curiosia from all earlier or other cultures. In other places, discussion about them is mingled with discussion of other string or stick records. Pieces of string with knots tied in them have been used by many people. Most of them were used as one-to-one tallies: that is, one sheep–one knot, two sheep–two knots, thirty-seven sheep–thirty-seven knots. Referring to Locke's conclusions but influenced, perhaps somewhat more, by Meade's foreword, in which quipus are compared with the tally cords of Bolivian shepherds, the ambiguity was continued. We find, for example, a discussion of quipus that emphasizes their importance and varied use in the Inca state but concludes with "We really do the same thing, if we want to consider knots we tie in a string or a handkerchief as a reminder."

Now, with a clearer notion of some of the ideas expressed through Inca pieces of string, we can more properly find the place of quipus in mathematics.

An attempt to do this, however, requires that we first look more closely at the nature and scope of mathematics. Mathematics arises out of, and is directly concerned with, the domain of thought involving the concepts of number, spatial configuration, and logic. In Western culture a professional class, called mathematicians, which traces its origins back to about 1600 in Europe, deals solely and exclusively with these issues. But they are not the only ones in our culture that utilize and are concerned with these concepts. Examples of other groups involved in mathematical endeavors are accountants, architects, bookies, construction engineers, landscape designers, navigators, and systems analysts. Nonprofessional mathematics, as practiced by members of these groups or by other individuals in our culture, may often be implicit rather than explicit. When the mathematical endeavors are implicit, they are, none the less, mathematics. Because of the provincial view of the professional mathematicians, most definitions of mathematics exclude or minimize the implicit and informal. It is, however, in the nature of any professional class to seek to maintain its exclusivity and to do this, in part, by recreating the past in terms of unilinear progress towards its own present. Moreover, involvement with concepts of number, spatial configuration, and logic, that is, implicit or explicit mathematics, is panhuman. In different cultures, it arises within different contexts. We can expect to find only implicit mathematics in cultures that do not have a professional class devoted to it. Since professional classes arise with urbanization and formation of the state, in cultures with these characteristics we can seek both implicit and explicit mathematics. The Inca state did have a professional class, called quipumakers, concerned with concepts of number, spatial configuration, and logic.

3 One distinction frequently heard in attempts at definitions of mathematics is that the numerical, geometric, or logical concepts are being explored and elaborated for themselves and not because of some other utilitarian end. While this, to us, is far too restrictive for an overall definition, it provides an excellent point of departure in looking at implicit mathematics outside of Western culture. There are many examples where mathematical concepts are related to practical or every day activities but where the elaboration far exceeds practical necessity. One simple example is mosaic tiling. In a culture where mosaic flooring is used, much attention may be focused on decorating the floor. The elaboration of geometric design can certainly exceed its practical function and become an area where, within that culture, an interest in geometric form, for itself, is being expressed. Another example is the logical structure of a particular kinship system. In some traditions, such as the contemporary West, the structure is minimal; in others, such as native Australia, it is a much more elaborate and dynamic element in daily life. A kinship system serves practical purposes, but we suspect that beyond a certain point, it is logical elaboration for itself. Games of chance and games of strategy may serve social or recreational purposes or may be

bet," *ISIS* 14(1930): 189–214. An example of a mathematical text which refers to quipus is Charles D. Miller and V. E. Heeran, *Mathematical Ideas* (Glenview, Ill.: Scott, Foresman & Co., 1968), pp. 3 and 25. The quotation is from Karl Menninger, *Number Words and Number Symbols* (Cambridge, Mass.: MIT Press, 1969), p. 254.

The need for viewing mathematics within a social and cultural context is also discussed in Dirk J. Struik, "On the Sociology of Mathematics" in *Mathematics Our Great Heritage*, ed. W. L. Schaef (New York: Harper and Bros., 1948). It originally appeared in *Science and Society* 6(1942): 58–70. Struik's article "Mathematics in Colonial America," pp. 1–7 in *The Bicentennial Tribute to American Mathematics 1776–1976*, ed. D. Tarwater (Washington, D.C.: Mathematical Association of America, 1977), although brief, is an example of a writing in the history of mathematics that includes the activities of practitioners with varied perspectives. Professional mathematics is viewed as a subculture in R. L. Wilder, *Evolution of Mathematical Concepts* (New York: John Wiley and Sons, 1968).

For a discussion of the origin of scientific specialization, see V. Gordon Childe, "The Bronze Age," *Past and Present* 12(1957): 2–15.

3 The Shongo children's game and the example from the Jokwe are from Claudia Zaslavsky, *Africa Counts* (Boston: Prindle, Weber and Schmidt, 1973). The entire book is highly recommended as it is one of the few books that present implicit mathematical ideas from other cultures. The Malekula example is from A. Bernard Deacon, "Geometrical Drawings from Malekula and Other Islands of the New Hebrides," ed. C. H. Wedgewood, with notes by A. C. Maddon, *Journal of the Royal Anthropological Institute* 64 (1934): 129–75.

For an introduction to graph theory and its history, see, Oystein Ore, *Graphs and Their Uses* (New York: Random House, 1963), and N. L. Biggs, E. K. Lloyd, and R. J. Wilson, *Graph Theory 1736–1936* (Oxford: Clarendon Press, 1976). The latter contains English translations of the original articles by Euler and Hierholzer.

associated with mythical beliefs or ritual practices. But within or beyond these domains, games can be places that basic mathematical concepts are expressed and elaborated. It is, therefore, necessary to look carefully at both nontechnological and technological aspects of a culture, for concepts that are implicit as well as explicit, and with an initially wide view of what constitutes ramification of the concepts of number, geometry, and logic.

We pursue the issue of comparison of mathematical ideas in different cultures through the use of an extended concrete case. The case is selected for several reasons. A primary reason is that, although it involves the solution of essentially the same problem in four different cultures, we can say with certainty that they in no way influenced each other. Secondly, the example shows an idea arising within very different settings within the different cultures. Also, the problem is nonnumerical. We will describe the problem in its different contexts, then relate them to each other, and then return to our discussion of mathematics in culture.

A Belgian, Emil Torday, lived for a while among the Shongo people in the Congo at the beginning of the twentieth century. He reported on a children's game that required drawing certain figures in the sand. An entire figure had to be drawn without lifting the finger and without retracing any line segment. One such figure is shown in figure 8.1. Can you draw the figure? (Hint: start at *A* and end at *B*.)

Fig. 8.1

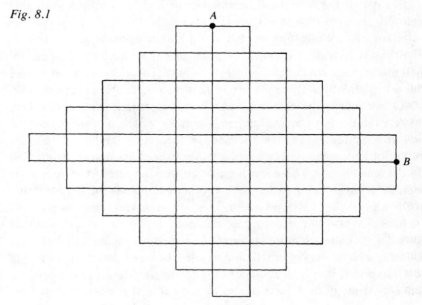

Among the Jokwe of Angola, there is a figure (fig. 8.2) illustrating the story of the beginning of the world. The sun is on the left; the moon is on the right; God is at the top; and man is at the bottom. Can you find the path that begins with God and returns to God without lifting your finger or retracing a line segment?

Fig. 8.2

A. Bernard Deacon, an anthropologist, collected 162 figures from the island of Malekula in the New Hebrides. Some are associated with religious beliefs, mythology, and rituals of the people, and others are secular. The drawing of the figures is handed down from generation to generation. According to one religious belief, ghosts of the dead must travel along a road to the land of the dead. The road passes a rock on which there sits a particular female ghost. At her feet is a diagram of the path to be traversed in order to get around the rock on which she sits. The guardian ghost always erases part of the diagram just as the ghost of the dead arrives. Those who do not know the proper path from memory can never safely arrive at the land of the dead. Many figures are associated with different aspects of this belief. One is shown in figure 8.3. Some of the figures are simpler and many of the secular figures are more complex. Again, try to trace the path without lifting your finger or retracing a line. This time, also try to memorize the diagram.

Fig. 8.3

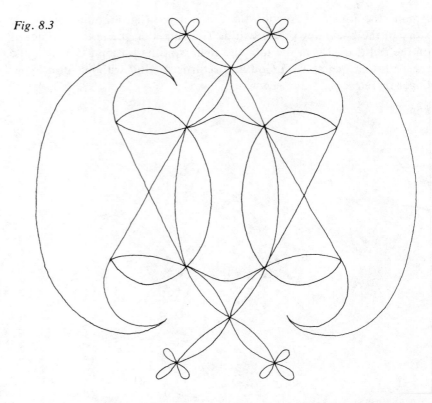

The river Pregel flowed through the city of Königsberg in Eastern Prussia. An island separated the river into two branches. There were seven bridges spanning the river so that people could travel from one part of the city to the other. In the early eighteenth century, people speculated about whether or not, on their Sunday strolls, they could find a way to walk around the city which would cross each bridge exactly once. In 1736, a mathematician, Leonhard Euler, decided to try to resolve the problem. He drew a picture similar to figure 8.4. The bridges are labeled a, b, c, d, e, f, g; the island is labeled A; and the banks of the river are B, C, D. He concluded that there was no way to walk around the city crossing each bridge only once. He also developed a general statement about similar paths that could or could not be traveled. Another mathematician, Carl Hierholzer, unfamiliar with Euler's work, wrote a paper in 1873 entitled "On the Possibility of Traversing a Line-system without Repetition or Discontinuity." (This is just a formal way of stating the question, Can a path be traced without lifting your finger or retracing a line?) Combining his conclusions with Euler's and using the terminology of current mathematicians, the Königsberg bridge problem is now restated.

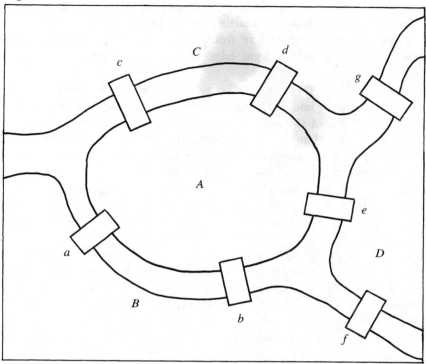

Fig. 8.4

The graph in figure 8.5 has vertices A, B, C, D (the land masses) and edges a, b, c, d, e, f, g (the bridges). Is there a path which traverses each edge of the graph once and only once? Such a path, if it exists, is now called an Eulerian path. Euler's and Hierholzer's conclusions were:

1. If a graph has exactly two vertices from which an odd number of edges emanate, an Eulerian path exists starting from either of these vertices.

2. If a graph has an even number of edges emanating from each vertex, an Eulerian path exists starting from any vertex.

3. In all other cases, no Eulerian path exists.

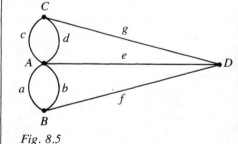

Fig. 8.5

Looking at the Königsberg bridge graph in figure 8.5, one can see an odd number of edges (three or five) emanating from each of the four vertices. Hence, no path exists. Notice that in the children's game in figure 8.1 there are exactly two vertices from which there emanate an odd number of edges—one is the point labeled A and the other the point B. Each of these vertices has three edges leaving it, and the other vertices each have four. Hence, there is a path and the figure can only be drawn with points *A* and *B* as the beginning and end. All the vertices in figure 8.2 have an even number of edges (four) emanating from them. The figure can be drawn by starting at any vertex and ending where you started. Finally, in the figure from Makula (fig. 8.3), every vertex is associated with four or eight edges and so an Eulerian path can be drawn.

So, whether in the context of games, religious beliefs, Sunday strolls in Königsberg, or line systems, people have pondered the same problem. Some emphasized the existence of a path and others emphasized the path itself. The Shongo children had some way, which we do not know, of arriving at their decision as to which figures could be drawn and, in the example we chose, they picked as starting and ending points the only two possible points for the entire figure. The Makula could elaborate their figures but still keep them traceable and, what is more, combined this with the requirement of memorization of geometric configurations.

Returning to our discussion of mathematics in culture, using this example, we now make several observations. In the Western history of mathematics, Euler's paper on the Königsberg bridge is considered the birth of the subject called graph theory. No predecessors or related events in other cultures are included. Clearly, these others did not inspire or effect Euler. The histories would be richer for the inclusion of these others, and the achievements of Euler and Hierholzer might even be more impressive because of the more universal nature of the question. We are not certain whether or not all Western mathematicians consider the ideas just included, from these other contexts in these other cultures, to be mathematics. But, assuming that they now would, what if graph theory did not yet exist? One book notes that because "it dealt largely with entertaining puzzles," the subject at first seemed mathematically "insignificant." Another calls the origins of the subject "humble, even frivolous." It would, we think, be all too easy to deny the humble and frivolous or religious or mythical from another culture if no professional Western mathematicians were yet working on the topic. And yet, because of the very difference in context, such ideas when coming from another culture, might even stimulate new avenues of approach.

4 The Inca state, as already noted, had a professional class involved in mathematical endeavors. When we look at their mathematical ideas, we are faced with limitations that apply when looking at ideas in any other culture. We are assuming that people in different cultures have some mathematical concepts in common. There are concepts in our culture that are not in theirs and, we can be sure, many in theirs that

4 For discussion of European mathematics in the fifteenth to seventeenth centuries see, for example, Carl B. Boyer, *A History of Mathematics* (New York: John Wiley and Sons, 1968), and Dirk J. Struik, *A Concise History of Mathematics*, 2d rev. ed. (New York: Dover Publications, 1948). Note, in

CODE OF THE QUIPU

are not in ours. Drawing on our own mathematical tradition, we may be able to see some concepts that are common. But, of necessity, even those are seen from our own mathematical and cultural framework and cast into that framework as we try to describe them. Many of their ideas will escape us. As such, even when we attempt to discuss them, it will remain, at best, a Western version of their ideas. Our study of quipus was limited by what we could see; our presentation of what we saw involved analogy and Western terminology. Essentially, our attempt was to render them understandable and yet not to make them appear a part of ourselves. We avoided using hypothetical Incan situations in our illustrative examples and exercises in order to emphasize that the descriptive framework is ours, and in order to minimize the implication that we truly understand how the concepts were integrated within the Incan context.

For the place of quipus in mathematics, we again draw largely on analogy with Western tradition, recognizing that we are limited in our knowledge of even that tradition because of what it has defined as its own history.

Clearly the quipus did not influence any practitioners of Western mathematics. The set of concepts cannot be said to be before, after, or at any particular time in Western mathematics. The numerical concepts included a base 10 positional system, arithmetic calculations, and ratios and proportions. Some of the quipumaker's numerical interests are combined with formal recording forms. These are akin to the interest in commercial arithmetic and bookkeeping of Europeans in the fifteenth to seventeenth centuries. For much of this, the Europeans were indebted to their trade with the Arabs. The linkage of numbers and formal recording forms are seen most clearly in the work of Fra Luca Paciolo and Simon Stevin. The former is considered the founder of modern accounting. Included in his book on double-entry bookkeeping, are the uses of arithmetic and algebra in trade and reckoning along with how to take an inventory and how to record based on that inventory. Stevin, noted earlier in relation to the European use of the decimal system, was influential in having double-entry bookkeeping used for and by government. We know that the quipumaker did political arithmetic. The term, meaning the collection, processing, and recording of statistical data of interest to the state, was first used in the late 1600s in England. The activity attracted the subsequent participation of such prominent mathematicians as Liebnitz and Laplace.

The Incan concepts of spatial configuration remind us more of ideas that came later in Western mathematics. In the first paper defining "trees" in 1857, A. Cayley, a noted mathematician, is interested in how many

particular, their discussions of the effect of trade and the role of the "reckon masters." An interesting facsimile of a seventeenth-century commercial arithmetic is Bernardus Saligancus, *The Principles of Arithmetic* (London, 1616; reprint ed., New York: De Capo Press, 1969). For discussion of the work of Paciolo and Stevins and the history of accounting see K. S. Johnston and R. G. Brown, *Paciolo on Accounting* (New York: McGraw Hill, 1963), and John B. Geijsbeck, *Ancient Double-Entry Bookkeeping* (n.p., 1914). For a discussion of statistics as political arithmetic, see Harold Westergaard, *Contributions to the History of Statistics* (London: P. S. King and Son, Ltd., 1932).

The article "On the Theory of the Analytical Forms Called Trees" by A. Cayley and an English translation of parts of the article by Vandermonde are also in *Graph Theory 1736–1936* cited above.

The Hindu use of color names for algebraic unknowns is discussed on pp. 75–84 of the 1974 reprint of Florian Cajori, *A History of Mathematical Notations* (1928; reprint ed., LaSalle, Ill.: Open Court Publishing Co., 1974).

differently shaped trees can be constructed with a given number of branches. For example, he shows that a tree of three branches can have three branches on one level, or two branches on the first and one on the second, or one on the first and two on the second, or one branch on each of three levels. But in, say, a tree with two branches on the first level, it depends on which branch the next level emanates from. He and a quipumaker might have had some good discussions. And, imagine the interest of a quipumaker in this first paragraph of an article written in 1771.

Whatever the twists and turns of a system of threads in space, one can always obtain an expression for the calculation of its dimensions, but this expression will be of little use in practice. The craftsman who fashions a braid, *a* net, *or some* knots *will be concerned, not with questions of measurement, but with those of position; what he sees there is the manner in which the threads are interlaced. It would therefore be useful to have a system of calculation more relevant to the worker's mode of operation, a notation which would represent his way of thinking, and which could be used for the reproduction of similar objects for all time.*

The article, by A.-T. Vandermonde, was entitled "Remarks on Problems of Position." This type of study of position grew into the field called topology. As for color, there is none in Western mathematics. While notation is of extreme importance, all notation retains the same meaning whether it is printed in black or written with white chalk. However, in the history of Hindu mathematics, we find that in the seventh century C.E. algebraic unknowns (for which we use x, y, z, etc.) were referred to by color names. The second through sixth unknowns in a problem were called black, blue, yellow, white, and red. Later, two-letter abbreviations of the color names were used for the unknowns instead.

The way the concepts of number, geometric configuration, and logic were formed together by the quipumaker was unparalleled in other cultures. Unfortunately, the intellectual endeavor of quipumaking came to an abrupt end during the sixteenth century. Although we may know where some similar concepts led in Western culture, since the Incas were not headed in the same direction, we cannot know where their ideas would have led.